DEUTERIUM DEPLETION

A New Way in Curing Cancer
and Preserving Health

Somlyai Gábor PhD

DEUTERIUM DEPLETION

A New Way in Curing Cancer and Preserving Health

Gábor Somlyai PhD

ISBN 978-615-01-4386-6

All rights reserved

© Gábor Somlyai 2022
© ATP Ltd.
Cover ©123RF
Translation Balázs Nagy 2021

Table of Contents

PREFACE — 15

Chapter One
GENERAL CHARACTERISTICS OF CANCER — 17
WHAT IS CANCER? — 17
ARGUMENTS SUPPORTING THE GENETIC THEORY OF CANCER — 20
ARGUMENTS SUPPORTING THE METABOLIC THEORY OF CANCER — 22
WHAT IS THE FOREFRONT NOW IN THE TREATMENT OF CANCER, AND WHERE IS IT HEADED? — 23

Chapter Two
A PARADIGM SHIFT IN BIOLOGY — 27
THE CONCEPT OF MOLECULAR AND SUBMOLECULAR BIOLOGY — 27
DEUTERIUM — 29
DEUTERIUM IN NATURE — 31
    Water with a lower-than-natural deuterium content: deuterium-depleted water (DDW) — 33
    The production of DDW — 34
THE BIOLOGICAL EFFECTS OF DEUTERIUM — 35
    Biological processes in an environment with a higher-than-natural deuterium content — 35
    Variation of the D/H ratio is a natural biological process — 36
    The effect of DDW intake on organisms' deuterium levels — 38
CELL PHYSIOLOGY FUNDAMENTALS — 39
    Cell membrane — 40
    Nucleus — 42
    The mitochondrion — 43

THE SUBMOLECULAR REGULATORY SYSTEM (SMRS)	45
The evolution of the regulatory mechanism	46
The submolecular regulatory system (SMRS) and the biochemical processes	47
The relationship between the submolecular regulatory system (SMRS) and genetic function	50
The submolecular regulatory mechanism and tumor necrosis	56

Chapter Three
CLINICAL RESULTS OF DDW APPLICATION IN CANCER PATIENTS 65

THE CONCEPT OF DDW DOSAGE	65
PROSPECTIVE PHASE II CLINICAL TRIAL ON PATIENTS WITH PROSTATE TUMORS	66
FOLLOW-UP (RETROSPECTIVE) STUDIES	68
Breast cancer	69
Prostate tumor	71
Pancreatic tumor	72
Lung tumor	72
Colorectal cancer	76
Results supporting the anti-cancer effects of deuterium depletion in the entire population	79
The use of deuterium depletion may prevent and protect against the recurrence of the disease	82
DYNAMICS OF DISEASE RELAPSE IN LIGHT OF THE RECENT RESEARCH	84

Chapter Four
LOW DEUTERIUM AS A KEY ELEMENT OF A HEALTHY LIFESTYLE 87

THE USE OF DEUTERIUM DEPLETION IN HEALTHY POPULATIONS	87
THE LIMITATIONS AND CONTRADICTIONS OF SCREENING TESTS	92

Chapter Five
USE OF DEUTERIUM DEPLETION IN BENIGN TUMORS 95

Chapter Six
DEUTERIUM DEPLETION AS A SUPPLEMENTARY TREATMENT FOR MALIGNANT TUMORS USED ALONGSIDE ONCOLOGICAL TREATMENTS ... 97
A PARADIGM SHIFT IN CURING CANCER ... 97
    A comparison of conventional and submolecular treatment strategies ... 99
    Deuterium depletion may influence tissue pathology ... 101
    The dilemmas of histological sampling ... 101
    The alignment of deuterium depletion with imaging tests, their sensitivity and influence on the results of the tests ... 106
    Deuterium depletion may affect tumor marker levels ... 108
    The correct definition of DdU and the principle of dosage ... 109

Chapter Seven
THE APPLICATION OF DEUTERIUM DEPLETION BEFORE DIAGNOSIS UP UNTIL THE START OF ONCOLOGICAL TREATMENTS ... 111
INDICATIVE SIGNS BEFORE DIAGNOSIS ... 111
THE DIFFICULTIES OF DIAGNOSTIC TESTS ... 113
THE APPLICATION OF DEUTERIUM DEPLETION IN BORDERLINE CASES WHEN IT IS REASONABLE TO SUSPECT CANCER, BUT NO DIAGNOSIS AND TREATMENT IS AVAILABLE YET ... 114
ESTABLISHING A DIAGNOSIS AND COMMUNICATING IT TO THE PATIENT ... 115
PLANNING A TREATMENT AFTER DIAGNOSIS ... 117
    Difficulties in planning conventional treatments ... 118

Chapter Eight
ADDITIONAL USE OF DEUTERIUM DEPLETION ALONGSIDE ONCOTHERAPIES ... 121
THE TIMING OF DEUTERIUM DEPLETION AND CONVENTIONAL TREATMENTS ARE DIFFERENT ... 121
    The conventional treatment according to the protocol has not started yet ... 122
        Chronic lymphocytic leukemia (CLL) ... 122

The patient has not consented to conventional treatment as prescribed by the protocol	123
Prostate cancer	123
Breast cancer	125
Lung cancer	126
Conventional treatment options have been exhausted	126
Stomach cancer	126
Liver cancer	127
Conventional treatments are completed	128
THE COMBINED USE OF SURGERY AND DEUTERIUM DEPLETION	128
The preoperative use of deuterium depletion	129
Picking the ideal time for a surgery	129
Head and neck cancers	130
Bladder cancer	130
Breast cancer	131
The postoperative use of deuterium depletion	131
THE COMBINED USE OF DEUTERIUM DEPLETION AND CHEMOTHERAPY	131
General advice for the complementary use of deuterium depletion alongside chemotherapy	132
The use of deuterium depletion following chemotherapy	133
The simultaneous use of chemotherapy and deuterium depletion	134
Deuterium depletion is started after or at the same time as chemotherapy treatment producing partial results	135
THE COMBINED USE OF DEUTERIUM DEPLETION AND HORMONE THERAPY	135
General guidelines for the combined use of deuterium depletion and hormone therapy	136
Hormone treatment prior to deuterium depletion	137
Deuterium depletion and hormone therapy are used simultaneously	137
Deuterium depletion should be started following a hormone therapy	138
SYNCHRONIZING DEUTERIUM DEPLETION AND RADIOTHERAPY	138
FITTING DEUTERIUM DEPLETION TO CONVENTIONAL TREATMENTS	139
Using deuterium depletion after surgery and alongside radiotherapy	140

The use of deuterium depletion during pre-operative radiotherapy
and following a surgery   141
    Rectal cancer   141
    Breast cancer   141
    Brain tumor (not glioblastoma)   142
Using deuterium depletion after a surgery alongside conventional
treatments   142
Using deuterium depletion combined with post-operative chemotherapy 143
    Adjuvant chemotherapy following a successful surgery 143
Using deuterium depletion alongside chemotherapy preceding a surgery 144
    Chemotherapy is used to achieve operability   144
Using deuterium depletion alongside chemotherapy and
(subsequently) radiotherapy   145
    The tumor is not operable, due to its location, staging, or classification 145
Using deuterium depletion alongside conventional treatments in
recently diagnosed stage III patients   146
The use of deuterium depletion combined with oncological treatments
in stage III patients receiving conventional treatments   147
Use of deuterium depletion alongside or following targeted therapies   147
    Herceptin/Breast cancer   148
    Gefitinib/Lung cancer   149
    Sutent/Kidney cancer   149
The combined use of deuterium depletion and immunotherapy   150

Chapter Nine
FACTORS AFFECTING DOSAGE AND THE EFFICACY OF DEUTERIUM DEPLETION, FINDINGS CONCERNING THE USAGE OF DEUTERIUM DEPLETION 153
    1. Daily DDW intake   153
    2. Deuterium concentration of DDW   153
    3. The body's response to deuterium depletion   154
    4. Body weight   155
    5. The type and histological classification of the tumor 155
    6. The tumor mass   156
    7. The shape of the tumor and the impact on the surrounding tissues 156

    8. The location of the tumor — 157
    9. Sensitivity of the tumor to deuterium depletion — 157
    10. Treatment of the primary tumor and/or metastasis — 158
    11. Classification of the tumor stage at the beginning of deuterium depletion — 158
    12. General physical condition of the patient — 158
    13. Other treatments — 159
    14. Complete blood count — 159
    15. Time elapsed since the start of deuterium depletion — 160

Chapter Ten
GENERAL ADVICE ON THE APPLICATION OF DEUTERIUM DEPLETION — 161

    1. Additional procedures that counteract deuterium depletion — 161
    2. How to consume DDW? — 162
    3. How does the deuterium concentration of DDW vary when boiled and kept in the open air? — 162
    4. On the carbonic acid content of waters — 163
    5. For how long should DDW be consumed? — 164
    6. Interrupting a deuterium depletion course — 164
    7. How to end a DDW course? — 164
    8. Long-term positive effects of deuterium depletion — 165
    9. Diets supporting deuterium depletion — 165
    10. Other additional procedures — 167

Chapter Eleven
THE MOST COMMON ACCOMPANYING SYMPTOM OF DEUTERIUM DEPLETION — 169

Weakness, prostration, drowsiness — 169
Increasing the dosage may also cause increased fatigue and drowsiness — 170
Blushing, increased temperature, and fever spikes — 170
Intermittently increasing pain — 170
Pain management — 170
Swelling and softening of the tumor-affected area — 171
Local warmth of the affected area — 171

Cerebral edema	171
Pulling and tingling sensation in the tumor	171
Minor bleeding in the bladder, stomach, or rectum	172
An improvement of appetite and general health	172
Weight gain	172
Exudation and wound healing in ulcerating tumors	172
An improvement of general comfort	172
Brick dust urine	172
Better tolerance of radiotherapy and cytostatic treatments	173
Transient coughing in lung cancer patients	173
Tumor necrosis may cause abscesses	173

Chapter Twelve
THE MAIN PHASES OF THE APPLICATION OF DEUTERIUM DEPLETION (1992–2020) — 175

Chapter Thirteen
A DEMONSTRATION OF THE EFFECTIVENESS OF DEUTERIUM DEPLETION THROUGH CASE STUDIES — 177

CASE STUDIES	177
Lung tumor	177
Breast cancer	184
Prostate tumor	188
Head and neck tumors	192
Colorectal cancer	193
Ovarian cancer	194
Cervical cancer	194
Melanoma malignum	195
Liver cancer	196
Grade I astrocytoma	197
Grade III astrocytoma	197
Glioblastoma	197
Neurofibromatosis	199

Bone marrow cancer	200
Myeloma	200
Acute myeloid leukemia (AML)	201
Chronic lymphocytic leukemia (CLL)	201

Chapter Fourteen
ADVICE ON ESTABLISHING THE DOSAGE

	205
RECOMMENDATION FOR HEALTHY PEOPLE, PREVENTION, ENHANCING PERFORMANCE	205
H/1 Protocol	205
H/2 Protocol	206
RECOMMENDATIONS FOR PEOPLE WHO ARE NOT YET DIAGNOSED WITH CANCER, BUT ARE EXAMINED FOR THE SUSPICION OF CANCER	207
P/D Protocol	207
RECOMMENDATIONS FOR CANCER PATIENTS	207
RECOMMENDATIONS FOR PATIENTS WHO HAVE RECOVERED TO PREVENT THE RELAPSE OF CANCER	208
RECOMMENDATIONS FOR PATIENTS WHO HAVE ACHIEVED A CANCER-FREE CONDITION WITH CONVENTIONAL THERAPIES TO PREVENT DISEASE RECURRENCE	208
C/R/1 Protocol	208
C/R/2 Protocol	210
RECOMMENDATIONS FOR PATIENTS WHO HAVE BECOME CANCER-FREE DURING THE COMPLEMENTARY USE OF DEUTERIUM DEPLETION TO PREVENT A RELAPSE	211
C/R/3 Protocol	211
RECOMMENDATIONS FOR CANCER PATIENTS TO ACHIEVE A CANCER-FREE CONDITION, TAKING INTO ACCOUNT THE CONVENTIONAL TREATMENTS USED	212
The patient is about to undergo surgery	212
C/C/Op Protocol	212
Inoperable patients receiving chemotherapy	212
I/C/C/Chem Protocol	212

Patients receive aftercare with adjuvant chemotherapy following
a successful surgery   213
   2/C/C/Chem Protocol   213
Recommendations for patients with glioblastoma, fitted to
the Stupp protocol   215
   3/C/C/Chem Protocol   215
Patients are inoperable and receive hormone treatment   216
   C/C/Horm Protocol   216
Patients receive radiotherapy   217
   C/C/Radther Protocol   217
SPECIAL ADVICE   218

Chapter Fifteen
**RECOMMENDATIONS FOR PATIENTS DIAGNOSED
WITH METABOLIC DISORDERS (M PROTOCOLS)**   221
   M Protokol   221

Chapter Sixteen
**RECOMMENDATIONS FOR ATHLETES
AND HEALTHY PEOPLE, TO ENHANCE PHYSICAL
PERFORMANCE**   223

APPENDIX
**SUMMARY TABLE OF THE DEUTERIUM CONCENTRATIONS
IN DIFFERENT NUTRIENTS**   227

BIBLIOGRAPHY   229

ACKNOWLEDGMENTS   237

DEUTERIUM DEPLETION *A New Way in Curing Cancer and Preserving Health*

Preface

I completed this book twenty years after my book, *Defeating Cancer!*, was published. Twenty years is a rather long period, not only in someone's life but also in the timeline of history and the history of science. In these two decades, many changes and events took place that we couldn't have even imagined at the end of the 1990s; it is highly likely that we will continue to face many unforeseeable challenges and changes now and in the future. The world was left speechless with the terror attack of the World Trade Center in New York in 2001, it survived a global economic crisis in 2008, was facing a significant migration crisis in 2015, and in 2020, the SARS-CoV-2 pandemic forced mankind to its knees and made it rethink its ways. The pandemic paralyzed the economy, streets became empty for months and the free movement, employment and pastime activities of people have been restrained in a way that has never been seen before. By the time this book was published, more than fifty million people had contracted the disease, resulting in the death of more than a million people. We don't know what the next twenty years will bring us, but in retrospect, we can already consider the changes that took place in curing cancer. Between 1999 and 2015, a 36% increase in new cases was recorded worldwide and in 2018, statistics showed 18.1 million new cases. It is predicted that by 2040, cases will increase to 29.5 million. This also implies that without significant changes to prevention and therapy, the number of those dying of cancer will grow from 9.6 million to 15.6 million in twenty years. The ongoing pandemic alerts the world's attention to the fact that we are facing important changes and decisions to preserve the environment and human life quality. This book aims to contribute to the better understanding and efficient treatment of cancer and other chronic diseases by introducing a new, submolecular approach that opens up a way for mankind to implement an efficient and sustainable therapeutic method devoid of harmful effects.

November 2020

Gábor Somlyai PhD

CHAPTER ONE

General Characteristics of Cancer

What Is Cancer?

Considering its many causes and clinical presentations, cancer is a complex medical condition, therefore it is hard to provide a brief, yet professional definition for this group of medical conditions. In a strictly professional sense, the expression "cancer" (carcinoma) only applies to malignant tumors, but people commonly use this expression to refer to any kind of malignant tumor. A common trait of malignancies is that a group of cells forms from a single cell during a given time (in a few weeks, but mostly in three, five, or up to ten years). This group of cells significantly differs from the surrounding healthy tissues in their functioning and morphology. A characteristic property is uncontrolled cell division, resulting in the tumor outgrowing its surroundings and spreading to the adjacent tissues. Cells breaking off of the tumor enter the blood and the lymphatic system to be carried further away in the body. Once they attach to the tissues there and continue growing, they create a new tumor (metastasis). In the case of hematopoietic malignancies, a mass of blastoid cells is released into the bloodstream.

On average, it takes four to five years until a single cell reaches a tumor of 0.5 to 1 centimeter in size that is detectable using current technology. By this time, the number of cells in the tumor is already over ten million. The "treacherous" nature of cancer can be traced back to this fact, as in the first four to five years the tumor stays below the detectable size threshold, with the patient experiencing no symptoms or complaints, during which time a great number of cells can break off of the tumor. If they manage to make it to other parts of the body, it takes another four to five years to create metastases, even if the primary tumor was removed years ago. (This is why we consider a five-year period to be crucial for cancer. If within these five years no new tumors appear in the body, then it is likely that the combined effect of conventional treatments and the body's immune system has destroyed all tumor cells.)

A tumor is therefore the "end product" of a complex, prolonged and multi-phase process. We can demonstrate this process with the example of a marble resting in a pit on a slope. Imbalances of the cell's genetic and metabolic processes occasionally cause "a loss of balance" which may dislodge the marble from its position, overcoming the minor obstacle holding the cell back to roll down the slope. Certain "fixing" mechanisms can help the cell find a way back to a safer point in the pit. In healthy cells, forces acting in the direction of the slope are in balance with the forces holding the marble back. The first event in the course of tumor formation is when a cell overcomes this obstacle. Once a cell "has broken loose", another pit may impede it on its way down, but if the forces acting in the direction of the slope outweigh others, it rapidly overcomes this obstacle as well. The more obstacles a cell "overcomes", the easier the next one will be to overcome, with the incline of the slope also increasing. If processes of imbalance cannot hold the cell back anymore, the cell goes tumbling down the slope and it is impossible to prevent it from starting an uncontrollable cell division.

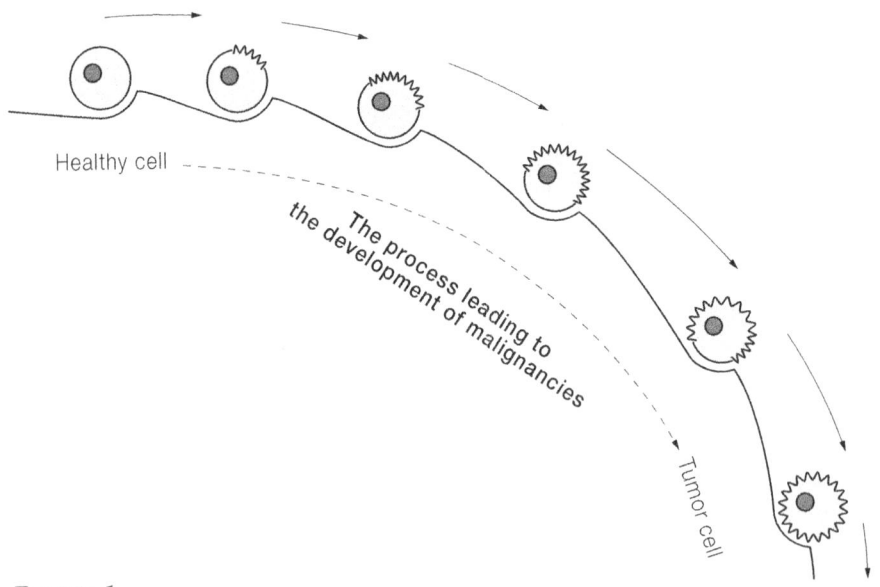

Figure 1

The multi-phase process of malignant cell degeneration, as demonstrated with a marble temporarily resting in a pit on a slope, on its path rolling down. It is apparent how a healthy cell becomes increasingly malignant during the years until nothing else stands in the way of uncontrolled cell division.

One generally accepted theory asserts that the primary cause of tumor formation is a sequence of errors in the genetic code [1]. During a person's lifetime, approximately 10^{16} cell divisions take place in the human body from fertilization to death. In every cell (apart from a few exceptions) of our body is a genetic code of 3.2 billion "letters". This genetic code is written with only four letters that correspond to the four bases constituting DNA, the cell's genetic material. These letters are the following: A for adenine, T for thymine, G for guanine, and C for cytosine. (This book contains approximately 360,000 letters, meaning that one can write the genetic code of a single cell in 9,700 books of the same length.) Cells duplicate their genetic code before each cell division so that the daughter cells have identical copies of their DNA. Enzymes (proteins) copying the genetic program make a mistake from time to time, inserting an incorrect letter for a given position in the genetic code of the daughter cell. If these errors occur at critical points of the 3.2 billion base pair-long genetic code which plays a key role in controlling cell division, then these cells will behave differently from the surrounding cells and will divide more frequently. This may lead to the group of cells outgrowing its surroundings, the macroscopic manifestation of which is a tumor.

The metabolic approach to cancer states that the underlying cause of cancer is a disruption in the cell's metabolism [2], primarily in the mitochondria, known as the powerhouse of the cell. This hypothesis was postulated by Otto Warburg [3] in the early 1920s, who subsequently received the Nobel Prize in Physiology in 1931 for his work. Adenosine triphosphate (ATP) molecules store the energy found in chemical bonds, resulting from the "burning off" of nutrients, in high-energy (so-called macroergic) bonds. The synthesis of ATP may take place in the presence of oxygen (aerobic metabolism) in mitochondria or without the presence of oxygen (anaerobic metabolism) in the cytoplasm. Terminal oxidation in the mitochondria (Szent-Györgyi-Krebs cycle) produces carbon dioxide and water as final products. Conversely, the complete oxidation of organic compounds does not take place in the cytoplasm. Along with carbon dioxide and water, lactic acid is released in the process, which in turn the cell utilizes for the synthesis of other molecules. Otto Warburg concluded that despite the availability of oxygen for tumor cells to facilitate the complete oxidation of nutrients in the mitochondria, the process takes place instead in the cytoplasm via anaerobic metabolism through glycolysis.

Arguments supporting the genetic theory of cancer

Evolution has created accurate repair mechanisms in cells to immediately fix errors resulting from incorrectly copying the genetic code. The sum of the number of errors occurring and the efficiency of the system used for error correction defines the "net" balance of genetic errors. This error-correcting system works with impressive efficiency. Approximately one error occurs when copying every one-thousandth letter in the DNA sequence. As a result of the DNA's repair mechanism, an entire copy of DNA contains less than one error per one million base pairs. In other words, a cell is incapable of repairing only one genetic error out of a thousand. The likelihood of tumor formation, therefore, depends on the frequency of errors occurring in the genetic code, and the accuracy (in percentage) of the cell's repair mechanism. (Note: Just as people are different in many aspects, everyone's repair mechanism is also different.) There are people whose repair mechanisms are very efficient, and there are others whose repair mechanisms work with a greater number of errors. This partly explains why one out of two people with an identical risk factor may fall sick, while the other does not.

Certain chemical substances significantly increase the risk of genetic errors. This explains why people handling carcinogenic, mutagenic, and toxic substances, active or passive smokers or those exposed to strong UV or radioactive radiation are at an increased risk of developing cancer. Is also easy to see that growth in the number of genetic errors increases the chance of something going wrong in the repair mechanism. Flaws in the repair mechanism may initiate a series of changes that can't prevent the cells, once out of control, from rapidly dividing.

In terms of prevention, everything stated above has two important messages: *(a)* every chemical substance that increases the risk of genetic errors occurring in the body also increases the likelihood of tumor formation; *(b)* the younger the age the first genetic errors occur, the earlier a cell starts to accumulate genetic errors and mutations that ultimately lead to tumor formation. This explains why smokers, people living in polluted areas, or those working with carcinogenic substances have a higher probability of developing cancer than the general population. As people get older, the chance of developing cancer [4] also increases (see Fig. 2). As more cell divisions take place in the body, the number of genetic errors also grows. Such growth also means that in a specific cell where a few genetic errors affecting the regulation of cell division have already occurred can easily push the cell to a point of no return.

People with a hereditary genetic predisposition to some diseases are also at a higher risk. Nowadays we possess the techniques to fairly accurately detect the genetic predisposition to cancer. For these people, it is extremely important to mitigate the risks and decrease the chance of further genetic errors and keep their bodies in a healthy condition.

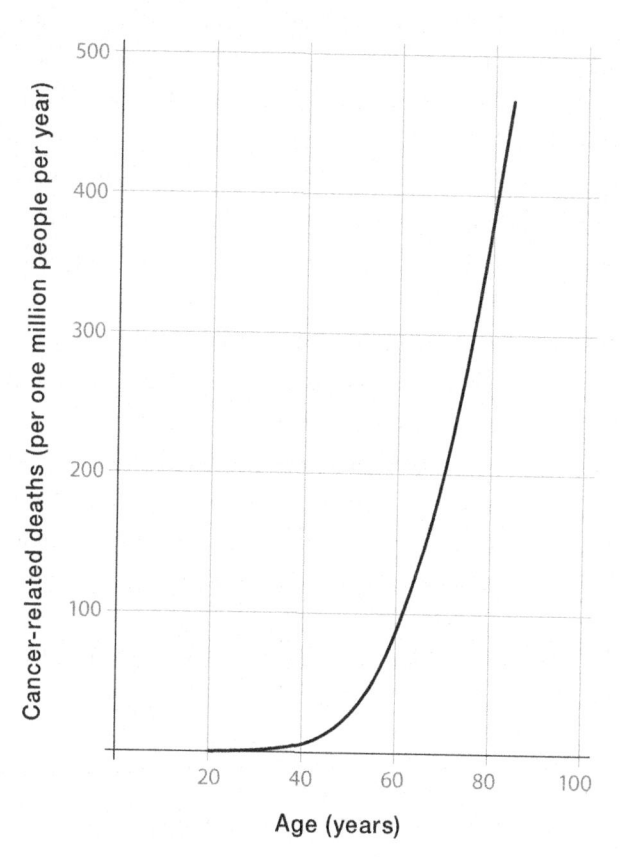

FIGURE 2

A graph of the number of those deceased as a result of cancer by age. The plot shows a surge in the death rate starting from age forty. In the population of people in their fifties, approximately thirty out of one million people die every year, whereas the same number is 400 for people in their eighties. (Source: [4] p. 1193)

General Characteristics of Cancer | 21

Arguments supporting the metabolic theory of cancer

The genetic approach states that the formation of cancer is caused by amassing many genetic errors. However, a contrary view states that the formation of clear cell renal cell carcinoma, a type of kidney cancer, for instance, is due to the mutation of one single gene in the mitochondrion, the gene encoding the enzyme fumarase hydratase [5]. Losing this genetic function also means that the TCA cycle (or Szent-Györgyi-Krebs cycle) of the mitochondria stops working. The cell is then unable to produce metabolic water which reduces the amount of deuterium. Research in which mitochondria have been transferred from tumor cells to healthy cells has demonstrated the key role of mitochondria. Once the mitochondria from the tumor cells were transferred to healthy cells, the latter also exhibited a tumor phenotype [2]. Contrary to expectations, when the nuclei of tumor cells were transferred to the cytoplasm of healthy cells, the cells remained healthy. Why do these cells not exhibit the characteristics of tumor cells even if their genetic material contains the mutations of tumor cells? The key role of metabolism is also suggested by epidemiological data showing significant variation in the incidence of certain cancers depending on the geographical region, lifestyle, and diet of the examined population group.

The secondary role of genetic errors is evidenced by the fact that the existence of genetic errors alone is not a sufficient condition for tumor formation. Genetic errors can be detected in many individuals, yet those people do not develop cancer. A good example is when in identical twins, one of the twins develops cancer and the other twin does not, despite their entirely identical genetic makeup.

I could go on and on, citing both supporting and opposing arguments [6] in favor and against each of these approaches, but it is difficult to come up with a final answer that reconciles the proponents of both ideas. I hope that the scientific evidence presented in this book and a new, submolecular approach to cancer resolves the apparent dispute between the two theories and show that both approaches are right in their own way.

What is the forefront now in the treatment of cancer, and where is it headed?

Different views may be made about the current situation in the treatment of cancer. These views depend on the professional background and considerations of those who voice them. The most important thing to make clear is that it is not justified to deal with cancer as if it were an incurable disease. Medical science has seen massive developments in the past few decades, making it possible for a significant proportion of patients with specific tumor types, such as testicular cancer or childhood leukemia, to be cured effectively with the proper treatment regimen. Even if a patient is not cured completely, a considerable breakthrough is that now it is possible to "tame" a chronic disease into becoming just an acute disease and to extend a patient's life while simultaneously maintaining his quality of life.

To provide a figure of how efficient modern therapies are, 9.6 million [7] cancer-related deaths were reported worldwide in 2018, with 33,000 in Hungary alone, even though the majority of patients received proper oncological treatment.

Given statistical data and trends, the question arises whether modern oncology is efficient. If we cannot face, accept and deal with the fact that further developing the current methods do not lead us to a real solution, then we cannot change the current situation. Embellishing and exaggerating partial results disregards today's available tools, instead of capitalizing on them. We need every new insight, every new area of development, and every new method that transcends the current treatment approaches and strategies. Using deuterium depletion is one such treatment strategy. To this day, nearly a hundred scientific papers have been published on the research and clinical results about deuterium depletion. This growing body of evidence supports the crucial role of naturally occurring deuterium in the regulation of biological processes.

In my book *Defeating Cancer!*, published in 1999, I quoted a few paragraphs of Tim Beardsley's editorial article from 1994 [8], reflecting on how President Nixon's war on cancer and the National Cancer Act of 1971 did not bring enough progress. That article's message is still relevant today. It quoted Dr. Peter Greenwald of the National Cancer Institute as being optimistic about how gene therapy, immunotherapy, and modifying the activity of specific genes would step up to the challenge. Reading this article from twenty-five years ago today, in 2020, it's apparent that no real breakthrough has been made in this time. While the therapies envisioned in the article have become available in the past twenty-five years, nevertheless we are not any closer to solving the real problem. Because today, cancer still beats us. In specific cases, new therapies have increased the life expectancy of the patients. However, this does not change the fact that every year, nine million people die of tumors (not to mention the sharp increase of treatment costs), and this number is expected to exceed thirteen million by 2030, according to the World Health Organization.

Massive databases are available on the incidence and survival rate of cancer, organized by tumor type, gender, and geographical distribution, etc. It is generally considered valid that 40 to 60% of detected new cases worldwide die of the disease (numbers vary across countries). On a global scale, 14 million new cases are detected, and 8.8 million people succumb to cancer. Detailed data sets are available, showing the number of new cases per year (morbidity) and the number of deaths per year (mortality). See Table 1 for these figures in the United States.

Type of cancer	Women			Men		
	morbidity	mortality	morbidity/mortality	morbidity	mortality	morbidity/mortality
Breast cancer	266,120	40,920	6.50	2,550	480	5.31
Cervical cancer	13,240	4,170	3.17	–	–	–
Colorectal cancer	47,530	23,240	2.04	49,690	27,390	1.81
Lung cancer	112,350	70,500	1.59	121,680	83,550	1.45
Stomach cancer	9,720	4,290	2.26	16,520	6,510	2.53
Endometrial cancer	63,230	11,350	5.57	–	–	–
Ovarian cancer	22,240	14,070	1.58	–	–	–
Leukemia	25,270	10,100	2.50	35,030	14,270	2.45
Esophageal cancer	3,810	3,000	1.27	13,480	12,850	1.05
Prostate cancer	–	–	–	164,690	29,430	5.59

Type of cancer	Women			Men		
	morbidity	mortality	morbidity/mortality	morbidity	mortality	morbidity/mortality
Liver cancer	11,610	9,660	1.20	30,610	20,540	1.49
Bladder cancer	18,810	4,720	3.98	62,380	12,520	4.98
Lip–oral cavity cancer	13,560	2,390	5.67	34,720	6,000	5.78

TABLE 1

New cancer cases (morbidity) and cancer-related deaths (mortality) in men and women in the United States in 2018, displaying the efficiency of therapies.

There's a morbidity/mo,rtality (the ratio of detected cases and deaths) column added to the American Cancer Society's chart. This column shows how efficient the current treatment for a specific cancer type is. This ratio is above 5 for breast, prostate, endometrial, and lip–oral cavity cancers. For cervical and bladder cancer, it is above 3. In women, the ratio is above 2 for three tumor types (colorectal cancer, stomach cancer, and leukemia). In men, the ratio is below 2 for colorectal cancer. The ratio is below 2 for all other tumor types (lung, ovarian, esophageal, and liver cancer), meaning that half of these patients die within a year. Tumor types for which the ratio is 5 or above respond well to treatment. A complete remission is possible, and the disease is curable when detected early on. Only a few tumor types belong to this group. For most cancers, the ratio is between 1 and 3. For lung cancer, the ratio is 1.45 in men and 1.59 in women, meaning that almost all patients diagnosed with lung cancer live hardly more than one year after the diagnosis. This figure is especially appalling, as lung cancer is one of the most common cancers.

In the past few decades, genomic research has been a trendsetter in the development of anti-cancer drugs. The development aims to detect and analyze genetic errors, with the ultimate goal of finding a tailor-made solution to cure cancer. While in recent years, new anti-cancer drugs have improved cancer death statistics and the life expectancy of the patients, no substantial breakthrough has been made. Developing these drugs has been and still is enormously expensive. Manufacturers incorporate these costs in the price of therapies. The monthly cost of an anti-cancer drug per patient is somewhere between $3,000 and $6,000. A disturbing fact is that while the cost of drugs and treatments multiply, minimal to no improvement is reported on efficiency.

Despite all our efforts, we are still facing the same challenge: can we make a substantial breakthrough in treating cancer?

CHAPTER TWO
A Paradigm Shift in Biology

The concept of molecular and submolecular biology

The areas of science concerning biological processes have their foundations on primarily the molecular view. In recent decades, chemistry and molecular biology research has revealed the biochemical and genetic processes in cells, explored biochemical pathways, and discovered the chemical reactions responsible for the integrity and basic functions of a cell. Research has also shown how biomolecules are formed and broken down and has mapped the structure of many proteins. In the modern era of drug development, researchers for decades have tested thousands of molecules to see which ones exhibit a positive physiological effect in a specific area of indication and whether the known biochemical processes can explain their mode of action. In modern-day oncology, it was the result of such tests that the effectiveness of a number of drugs against tumors was verified, and the drug was registered. Even despite the partial results, it is clear that there still isn't an effective cure for cancer. Identifying the first oncogene was an initial breakthrough. Molecular genetics research has enabled targeted drug development. However, this approach still worked on the molecular level, targeting the protein encoded by a specific gene. It has been generally accepted that with the new approach, a real and palpable breakthrough is within reach and is only a few years ahead. We have identified hundreds of genes associated with cancer formation, and there are hundreds of currently ongoing clinical research projects, yet we are still waiting for a truly effective cure.

The important question to ask – and we should have asked this question decades ago – is whether the regulation and harmonizing of the rapid and complex biochemical, genetic, and cell physiology processes can take place on a molecular level. Albert Szent-Györgyi was ahead of his time, as he asked the same question more than forty years ago [10, 11]. His thinking was that large protein molecules (usually these are the main targets of drug development) cannot fulfill this role of regulating the rapid, complex processes that take place in cells. Szent-Györgyi reasoned that if two molecules are composed of identical atoms, with one of them having one electron more or less, then these molecules differ on a submolecular level, and this fact may be crucial with respect to the regulation of cellular processes. He stated that this subatomic particle (the electron) is able to fulfill the role of regulation due to its small weight and mobility. Electrons are capable of regulating and synchronizing molecular processes. If their free movement is disrupted, it can lead to cancer.

After watching Szent-Györgyi's famous television interview in the 1970s, I continued thinking about submolecular regulation. I formulated the idea that perhaps it's not the negatively charged electrons, but the positively charged hydrogen ions that have a key role in regulating cell division and thus the formation of cancer. I held this idea with such certainty that from then on, I reviewed and reflected on all scientific knowledge based on it. Four years later, as a student of biology at the József Attila University at Szeged, Hungary (today's University of Szeged), I was preparing for my exam in physical chemistry. When learning about enthalpy, entropy, and pH, all of a sudden, I realized that deuterium, a naturally occurring heavy isotope of hydrogen and the other isotope of hydrogen *together* may fulfill a key role in the functioning of genes and enzymes, and in regulating cell division [12]. Following a career detour in the field of molecular biology, earning a doctoral degree, and completing two scholarships abroad, I waited altogether ten more years to finally begin to conduct experiments to prove the existence of a submolecular regulatory system (SMRS).

To validate my hypothesis, I only needed to ask one simple question: "Does naturally occurring deuterium play a role in regulating physiological processes?" The easiest way to provide an answer to the question is to examine whether a lower than natural deuterium concentration (deuterium depletion) changing the deuterium/hydrogen ratio affects physiological processes in the cells and living organisms.

Results of our research conducted in the past thirty years [12, 13, 14, 15, 16, 17, 18, 19, 20, 21], [22, 23] and independent research results [24, 25, 26, 27, 28, 29, 30] confirmed that deuterium depletion changes a number of cellular processes in different biological systems. Observations indicate that cells sense a change in the concentration of deuterium, and these changes induce and influence several processes in the cells and other organisms.

Results hint at the existence of a submolecular regulatory system, evolved during the billions of years since the appearance of life. This system uses the deuterium/hydrogen ratio to regulate fundamental genetic, biochemical, and physiological processes.

Deuterium

Hydrogen has three naturally occurring variants (isotopes): ^{1}H (hydrogen, H), ^{2}H (deuterium, D), and ^{3}H (tritium, T). Deuterium is a stable, non-radioactive isotope of hydrogen, having a proton and a neutron (of the same weight) in its nucleus (see Fig. 3). (The nucleus of tritium contains two neutrons alongside a single proton. This makes the nucleus unstable and causes this isotope of hydrogen to be radioactive.) It has been known for decades that as a result of the weight difference between H and D, molecules containing deuterium behave differently in chemical reactions. For instance, if a chemical bond contains deuterium instead of hydrogen, it takes six to ten times more time to split the bond during a chemical reaction [31, 32, 33]. The first attempt to produce heavy water (D_2O) was also based on this observation. During the electrolysis of water (as a result of an electric current, hydrogen and oxygen molecules are released from the water in a chemical reaction), the dissociation speed of H_2O exceeds that of D_2O by several times. In the remaining water, the concentration of heavy water steadily increases. Substituting hydrogen for deuterium somewhere else in the molecule, and not in the chemical bond that is about to split, substantially slows down chemical reactions. This so-called kinetic isotope effect offers a rare insight into how chemical reactions work, and why substituting hydrogen for deuterium is a widely used method in chemical research.

Electron Weight: 1/1840
Proton Weight: 1

Hydrogen (H)
Weight: 1

Electron Weight: 1/1840
Proton Weight: 1
Neutron Weight: 1

Deuterium (D)
Weight: 2

(H) (H)
8 electrons
8 protons
8 electrons
Oxygen atom
Weight: 16

H₂O
Water
Weight: 18

(H) (D)

HDO
Semi-heavy water
Weight: 19

(D) (D)

D₂O
Heavy water
Weight: 20

Tap water
In a glass of tap water, there is 1 deuterium atom for every 6600 hydrogen atoms or 150 D₂O (150 ppm) molecules for every 1 million H₂O molecules

Deuterium-depleted water
In a glass of deuterium-depleted water with deuterium 25 ppm concentration, there is 1 deuterium atom for every 40,000 hydrogen atoms and 25 D₂O molecules for every 1 million H₂O molecules

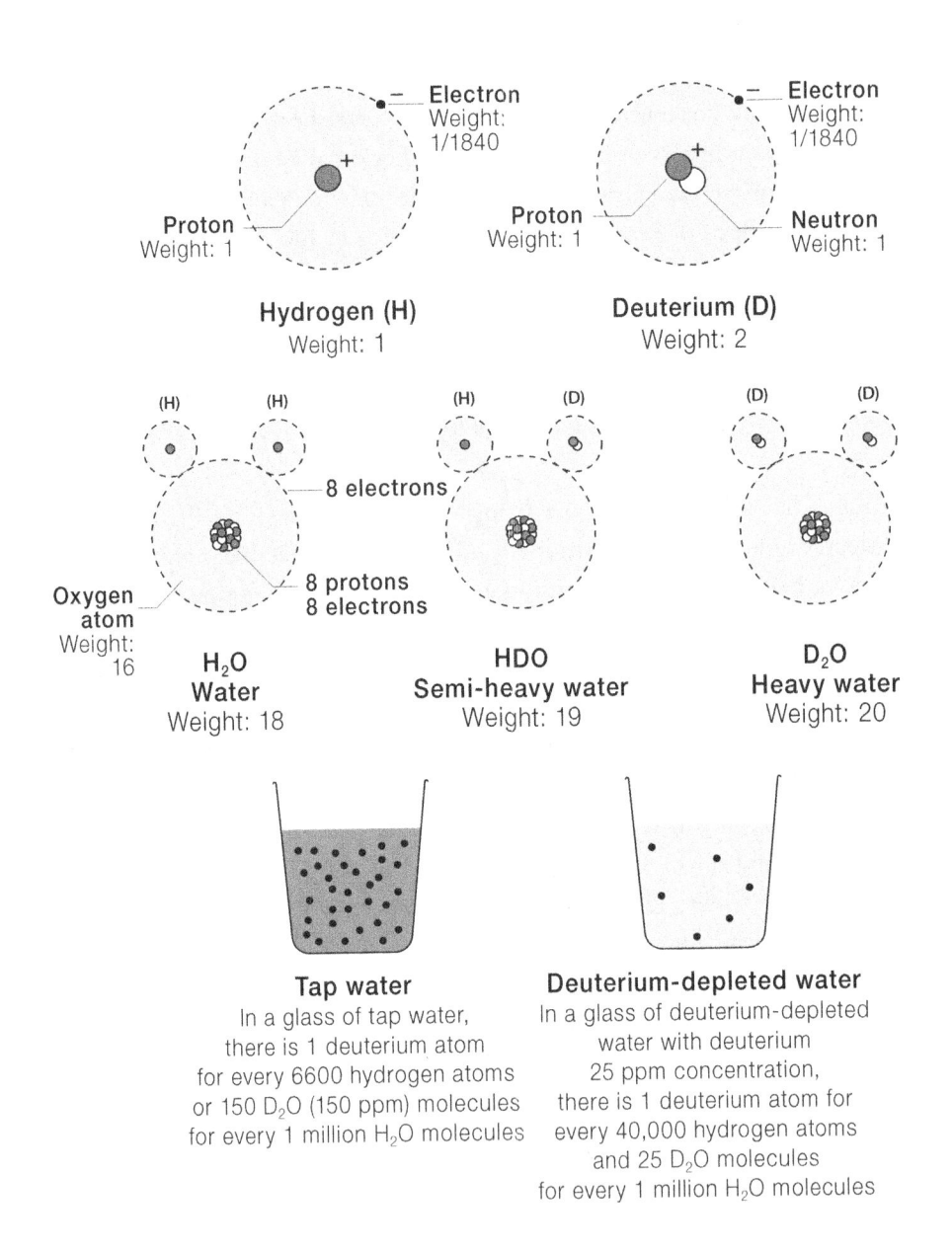

FIGURE 3

Hydrogen, the simplest chemical element, consists of a positively charged proton and a negatively charged electron. (Its atomic weight is 1). A deuterium's nucleus is made up of a proton and a neutrally charged neutron of equal weight. This fact causes a 100% weight difference between the two stable isotopes of hydrogen. In terms of the D/H isotopes, a glass of water contains three types of water molecules: light water (H_2O), semi-heavy water (HDO), and heavy water (D_2O). In deuterium-depleted water, the number of deuterium-containing molecules is smaller, therefore the D/H ratio substantially shifts towards hydrogen.

Deuterium in nature

On our planet, the deuterium content of living organisms is primarily determined by the deuterium content of evaporating ocean water, which after atmospheric circulation returns to the surface of the Earth in the form of precipitation such as rain and snow.

When evaluating deuterium measurements in precipitation at hundreds of points on Earth [34], we can conclude that the deuterium content of precipitation decreases from the Equator to the North Pole and the South Pole, from the oceans to the inland, and decreases with higher altitude above sea level. The difference in vapor pressure between H_2O and D_2O (also HDO) explains this observation. (As a matter of fact, the so-called fractional distillation process used in some nuclear reactors to produce heavy water is also based on this difference.)

The measurement unit ppm (parts per million) is used to determine the deuterium content of deuterium-depleted water (DDW). The number of ppm units shows how many deuterium (D) atoms there are out of one million hydrogen atoms (H). It also shows how many heavy water molecules (D_2O) there are out of one million water molecules (H_2O). (Note: In surface waters, deuterium is usually not present in the form of D_2O, but as HDO.)

In the temperate climate zone, the deuterium content of surface waters is 143–150 ppm (that is to say, out of one million hydrogen atoms, there are 143–150 deuterium atoms) with minimal fluctuation. Contrarily, around the Equator, the deuterium content of surface waters is 155 ppm, and 135–140 ppm in inland Northern Canada (see Fig. 4).

Figure diagram labels:
- As we move towards the poles, the deuterium content of precipitation decreases
- 130 ppm
- As we move towards the poles, the deuterium content of precipitation decreases
- 155 ppm
- 130 ppm
- The deuterium content of precipitation decreases with altitude
- Water evaporating from the oceans
- As we move further inland from the sea, the deuterium content of precipitation decreases

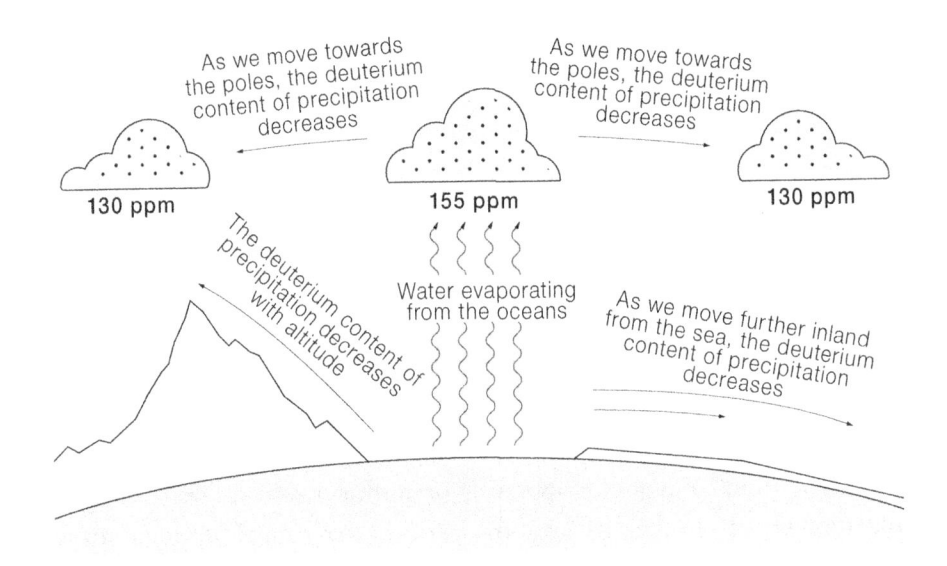

FIGURE 4

While the deuterium concentration of precipitation samples from different parts of the Earth varies, the reason for this is due to the different physical properties of H_2O, HDO, and D_2O. In the equatorial regions, the deuterium content of water evaporating from the oceans is approximately identical to the deuterium concentration of ocean water (154–155 ppm). However, as clouds move to the north or the south and release their water content in the form of rain and snow, HDO and D_2O molecules precipitate from the clouds at a higher rate than they occur in the cloud. Closer toward the arctic regions, clouds gradually release their heavy water content. The same phenomenon occurs when clouds are forced over physical barriers, such as high mountains (the heavy water content of precipitation at high altitudes in the mountains is lower than that of the precipitation at the foot of the mountains). The heavy water content of precipitation also decreases further inland, as compared to coastal regions.

If we calculate the 150 ppm value to a molar concentration (mmol/L), it becomes apparent that in natural waters, the concentration of D_2O is 8.4 mmol/L, corresponding to a 16.8 mmol/L HDO concentration. Deuterium is predominantly found in natural waters in the form of HDO.

Approximately up to 60% of the adult human body is water. Adjusting the above concentration with this value, and taking into account that other organic and inorganic compounds may contain deuterium, we should calculate that the total deuterium concentration in the human body is approximately 12 to 14 mmol/L. (To demonstrate the quantity of deuterium in the human body, we can give an estimate for the deuterium and hydrogen content of a body weighing roughly fifty kilograms [110 pounds]. A body of this weight would contain five kilograms [11 pounds] of hydrogen and 1.5 grams [0.053 ounces] of deuterium.)

To compare, human blood serum's concentration of calcium ranges between 2.24 and 2.74 mmol/L, magnesium between 0.75 and 1.2 mmol/L, and potassium between 3.5 and 5.0 mmol/L. Thus, the concentration of deuterium is six times greater than the concentration of calcium, ten times greater than that of magnesium, and three times greater than that of potassium. Considering the above concentrations, we might ask the logical question: If the presence of elements such as calcium, magnesium, and potassium in a much lower concentration than deuterium is indispensable for life, how could science disregard deuterium's significance for over sixty years?

Water with a lower-than-natural deuterium content: deuterium-depleted water (DDW)

Deuterium-depleted water (DDW) is a type of artificially produced water with a composition different from natural waters. In natural waters in the temperate climate zone, on average 143 to 148 out of one million water molecules (H_2O) are so-called semi-heavy water molecules (HDO) and heavy water molecules (D_2O). The deuterium concentration in the Danube River is 143 to 144 ppm. Summer precipitation can have up to 154 ppm deuterium concentration, with winter precipitation as low as 130 ppm. The DDW concentrations used in my experiments have ranged between 25 and 125 ppm. (Deuterium in this low concentration range is present almost exclusively in the form of HDO.)

The production of DDW

The most common method of producing DDW is so-called fractional distillation. The separation process is based on the different boiling points of H_2O, D_2O, and HDO. Water molecules containing deuterium have a 1.5°C higher boiling point. Consequently, if water is evaporated, the concentration of D_2O is a few percent lower in the resulting vapor (as it contains less deuterium). For every boiling/condensation cycle at 100°C the deuterium concentration in the water vapor decreases by approximately 1–1.5 ppm. Once the resulting water vapor is condensed and evaporated again, a further decrease of deuterium concentration in the water vapor can be achieved. On an industrial scale, this process takes place in ten- to thirty-meter-high (thirty to one-hundred feet high) distillation towers. Depending on the size of the tower, hundreds of kilograms (or even up to a thousand kilograms, that is, a ton) of vapor per hour is fed into the tower. By the time the vapor reaches the top of the tower, it has condensed and evaporated thirty to one-hundred times, resulting in just a few tens of liters (a few gallons) of DDW. The end product can be regarded as ultra-clean water, devoid of all the dissolved salts, contaminations, and parts of the semi-heavy and heavy water molecules.

(Several websites state that certain procedures or household devices are allegedly suitable for effectively reducing the concentration of deuterium.) In some of these cases, the principle does work, yet the devices/procedures are not suitable for effectively reducing the deuterium concentration of water, lest to speak of satisfying the daily DDW intake of a person. The journal *Clinical and Experimental Health Sciences* refutes false views which deem it possible to significantly reduce the deuterium concentration of water using only DIY cryogenic methods. The cited research used nineteen liters (five gallons) of water, frozen in approximately 200 mL (6.7 ounces) doses. After two to three hours, the frozen parts were disposed of, keeping only the liquid water. Researchers repeated these steps with the remaining water, each time measuring the deuterium concentration of the initial product and the end product. Measurements demonstrate that up to two hundred milliliters (six ounces) of water produced by this laborious process had a deuterium concentration of 144 ppm. This was only 3 ppm lower than the initial value of 147 ppm. Results provide empirical evidence to refute the idea of possibly reducing the deuterium concentration of water by using this method [35].

The biological effects of deuterium

Biological processes in an environment with a higher-than-natural deuterium content

The different chemical behaviors of deuterium and hydrogen also manifest in biological systems. In the years following the discovery of deuterium (from the 1930s on), researchers observed that applying heavy water in a large concentration has a substantial effect on the processes in a biological system. These experiments proved that, for instance, the growth rate for tobacco plants differs depending on the percentage of heavy water used for watering the plants. When given normal water, plants experienced ordinary growth rates. When the amount of heavy water increased, the growth of the tobacco plants was inhibited, developing shorter and shorter shoots [36]. Substantial changes were registered in the case of the mold *Aspergillus niger*. *Aspergillus niger* – as its Latin name implies – is a black mold that retains a soft, creamy white color when grown in heavy water. That is, in this medium, the mold cannot produce the pigment responsible for its black color. One of the simplest living organisms, the *Euglena* species (single-cell flagellate eukaryotes containing chlorophyll and sensitive to red light), show a reaction to light by day but show no reaction to light at night. If placed into water with a deuterium concentration of more than 45%, their ability to react to light stops. Replacing them into light water allows them to regain their ability, but their circadian rhythm changes. The longer in water with high deuterium concentration, the more their circadian rhythm changes. Among all other known environmental and chemical factors, only deuterium has the ability to change an organism's circadian rhythm. Green algae, contrarily to other organisms, draw the hydrogen necessary to synthesize organic compounds from water only. Green algae grown in heavy water are (except for their physical appearance) different in many aspects from algae grown in ordinary water. Their composition of proteins, carbohydrates, and nucleic acids is different, and photosynthesis in these algae is three times slower than normal. Compounds from these algae, rich in deuterium, are used in important biological studies. Heavy water has a detrimental effect on animals' blood counts, and in extreme cases, heavy water with a concentration higher than 35% caused the death of the animal (dog) [37]. Studies conducted with mice and rats also confirmed that in complex organisms, as opposed to simpler organisms, hydrogen cannot be fully substituted for deuterium.

The effect of heavy water on living organisms is not surprising considering that the substantial makeup of organisms is water. Heavy water differs from ordinary water (H_2O) in many aspects, including its chemical properties. The melting point of heavy water is 4°C higher, its boiling point is 1.5°C higher, its density is 10% higher and its viscosity is 25% higher than that of light water. All these supports the generally accepted view that the structure of heavy water is "more robust" than that of ordinary water. A part of the hydrogen present in living organisms – hydrogen atoms connected to oxygen, sulfur, and nitrogen – is quickly replaced with deuterium in heavy water. This way, the hydrogen bonds stabilizing the structure of proteins and polypeptides are substituted for the stronger deuterium bonds. This explains why the structure of proteins is more stable in heavy water, and more resistant to denaturation and conformation changes.

Variation of the D/H ratio is a natural biological process

Science and measurement techniques have made it possible to detect the presence of deuterium and to exactly determine its distribution in the molecules constituting a cell, in nutrients, and the body. It has been shown that in specific plant molecules, the ratio of deuterium and hydrogen may substantially differ from the environment. Metabolic processes in plants explain this difference. If a plant, for instance, uses the so-called C3 or C4 biochemical pathways to fix carbon from the atmosphere, the concentration of deuterium decreases to different extents in sugar molecules. Plants using CAM photosynthesis may, under certain circumstances, subsequently raise their deuterium concentration. This means that deuterium concentration in the human body is substantially affected by the plants in our diet. For example, spinach, wheat, rice, and barley use C3 carbon fixation. Maize, sugar cane, millet, and sorghum use C4 carbon fixation. In the sugar molecules of C3 plants, the deuterium concentration is 10–15 ppm lower than that of C4 plants [38, 39].

In algae, cells discriminate between hydrogen isotopes in processes in light but don't make the same distinction in darkness [40]. This attests to the intricate nature of physiological processes and to their sensitivity to changing deuterium/hydrogen ratios.

Other significant differences were detected when examining the deuterium concentration of various foods. Table 2 shows the deuterium concentration in the dry matter of different nutrients. These values correspond to the deuterium concentration of 0.3–0.4 liters (ten to thirteen ounces) of metabolic water produced in the mitochondria.

Flour	150 ppm
Sugar	146 ppm
Cottage cheese	136 ppm
Olive oil	130 ppm
Butter	124 ppm
Pig fat	118 ppm

TABLE 2

Deuterium concentration in the dry matter of different nutrients

Studies examining the distribution of deuterium in organic compounds have proven that the location and distribution of deuterium within the molecules are not random but determined and characteristic of a specific molecule. Studies have also shown that there is a substantial difference between the deuterium concentrations of carbohydrates and animal fats. In linolenic acid, a fatty acid-containing eighteen carbon atoms, the deuterium concentration on the ninth carbon atom is only 60 ppm. It then is 120 ppm on the tenth carbon atom, 90 ppm on the eleventh, 120 ppm on the twelfth, and 60 ppm on the thirteenth [41, 42]. Similarly, research has demonstrated that the distribution of deuterium in carbohydrates deviates from the statistically expected random distribution and is specific to a given plant [43]. These observations suggest that the isotope effect in biochemical processes and the deuterium concentration of the various molecules involved in chemical reactions determine the occurrence of the heavy isotope of hydrogen, deuterium, in a specific location of the compounds. The distribution and location of deuterium in different molecules and the resulting changes are therefore determined and not random. This also indicates that the isotopes; deuterium and hydrogen play a regulatory role in living organisms.

The effect of DDW intake on organisms' deuterium levels

A decrease in deuterium levels as a result of DDW was proven in animals and humans. Using a veterinary formula with 25 ppm deuterium (Vetera-DDW-25®) reduced the deuterium content of canine blood serum from 145 ppm to 65 ppm in three months. In a Phase II human clinical trial, thirty patients diagnosed early on with diabetes drank 1.5 liters (fifty ounces) of DDW with a concentration of 104 ppm a day for ninety days. At the beginning of the clinical study, the average deuterium concentration in the blood serum was 147.5 ppm, ranging between 146–150 ppm among the patients. By the end of the third month, the average deuterium concentration decreased to 133.9 ppm (p ⊠ 0.0001). At this point, deuterium concentration in the serum ranged between 125 and 143 ppm. In terms of individual differences, the smallest decrease in the concentration of deuterium was 4 ppm, while the greatest decrease was 24 ppm [44]. This demonstrates that besides the concentration of DDW consumed, a range of other factors may contribute to the scale of decrease. The patient exhibiting the smallest decrease (4 ppm) drank less than the prescribed daily amount of 1.5 liters (fifty ounces) of DDW (104 ppm), and/or consumed a substantial amount of other liquids with normal deuterium concentration. Figure 5 shows how the deuterium concentration changed in the blood of the patients participating in the clinical study [44].

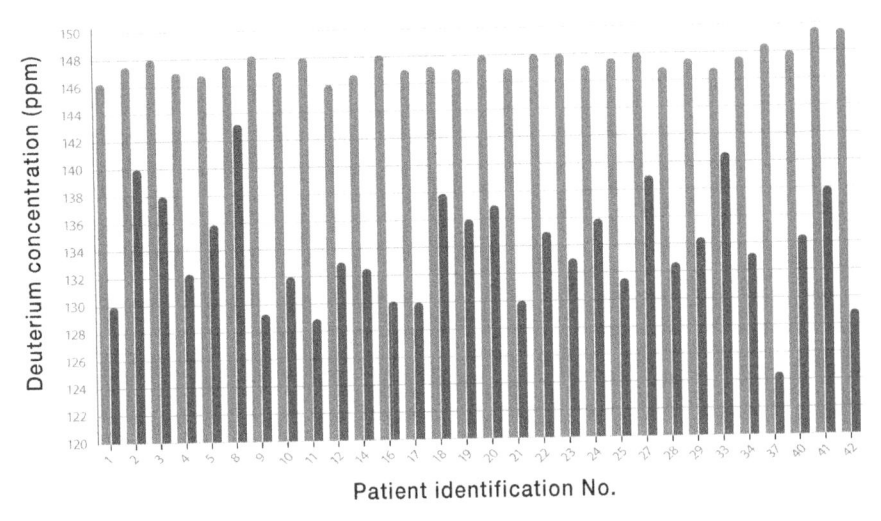

FIGURE 5
Deuterium concentration in the patients' blood serum at the beginning of the study (Day 0) and on the ninetieth day of the study.

Serum deuterium concentration in other patients who consumed DDW in different concentrations for months established a clear relationship between the average daily intake of deuterium (DDW and nutrients) and the deuterium concentration of the blood (see Table 3). To cite an example, the greatest detected decrease of 82 ppm (150 ppm − 68 ppm = 82 ppm) was achieved as a result of consuming DDW with a concentration of 25 ppm.

Average deuterium concentration of daily intake (DDW and nutrients)	Deuterium concentration of blood serum (± 1 ppm)
41 ppm	68 ppm
76 ppm	102 ppm
88 ppm	110 ppm
94 ppm	118 ppm
108 ppm	136 ppm

Table 3

Ingested water and nutrients together determine the deuterium concentration of the blood.

Cell physiology fundamentals

Before diving into a detailed introduction to submolecular regulation, it is worth first explaining the elements of a cell that regulate the cellular processes. Readers might find this overview useful to have a grasp of how cells and the submolecular regulatory system work.

To demonstrate the size of an average cell, picture a ruler with both the inch and metric scale. Now, if you take a one-millimeter increment (one-tenth of a centimeter), that corresponds roughly to one twenty-fifth of an inch. If you break this millimeter increment further down into twenty-five parts, that equates to forty micrometers, which is how big an average cell is. Figure 6 shows a schematic drawing of a cell, demonstrating its key constituents, the organelles. For these constituents is provided a short description about the cell membrane, the nucleus and the mitochondrion, as these three elements play a key role in the submolecular regulation of cellular processes.

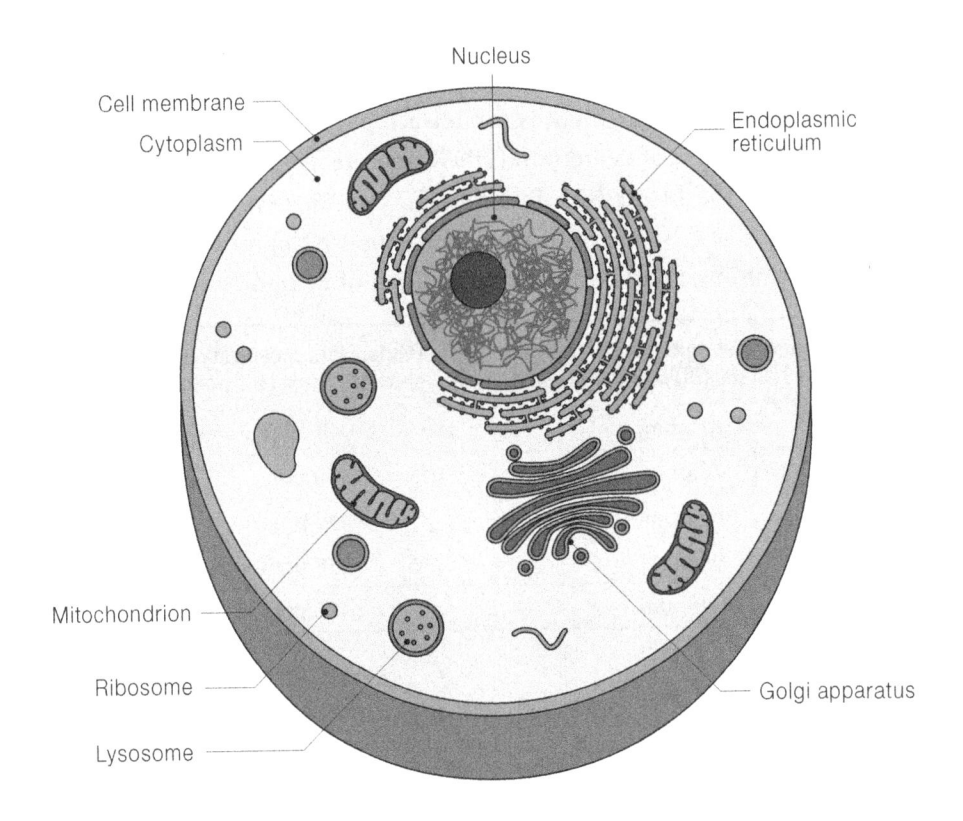

FIGURE 6
Schematic diagram of a eukaryotic cell

Cell membrane

Cells are enclosed by a cell membrane (see Fig. 7 for a schematic diagram). Imagine the cell membrane as a thin layer of fat (it is five millionth of a millimeter thick) in which proteins float around like icebergs in a sea. Some proteins go through the membrane, some are only attached to one side of the membrane. The cell membrane separates, but also connects the cell to its surroundings. Proteins in the cell membrane each fulfill a specific role. A particular group of proteins, for example, ensures the proper concentration of sodium, potassium, magnesium, and calcium in the cell. As a result, one of the transport systems ensures a higher concentration of potassium and a lower concentration of sodium inside the cell, while maintaining higher sodium levels and lower potassium levels outside of the cell. Another group of proteins acts as a receptor. They sense when a growth hormone binds to the membrane and triggers a signal to the other side of the membrane, ultimately resulting in a chain reaction inside the cell. A pathological examination is crucial to establish the hormone status in the case of a breast tumor. In patients whose tumor cells have such receptors in their membranes, it is possible to use drugs that block the hormone from binding to the receptor in the membrane and thus prevent the tumor from growing. The sodium–proton exchanger (Na^+/H^+) is an important transport process that takes place in the cell membrane. Once activated, the sodium–proton exchanger takes up sodium (Na^+) from the cell's surroundings and transports a hydrogen ion (H^+) out of the cell. As you will see later, this process is of special importance. The growth hormone stimulating cell division is capable of activating this transport process, shifting the cell's pH to alkalinity (and reducing the concentration of H^+). It also alters the deuterium–hydrogen ratio in the cell, as the transport protein is "picky," and removes only the "light" isotope from the two hydrogens. The cell membrane should be considered a part of the cell that separates the cell from, but one that also connects it to its surroundings, as the "sensors" on the membrane's outer surface continuously send signals into the cell.

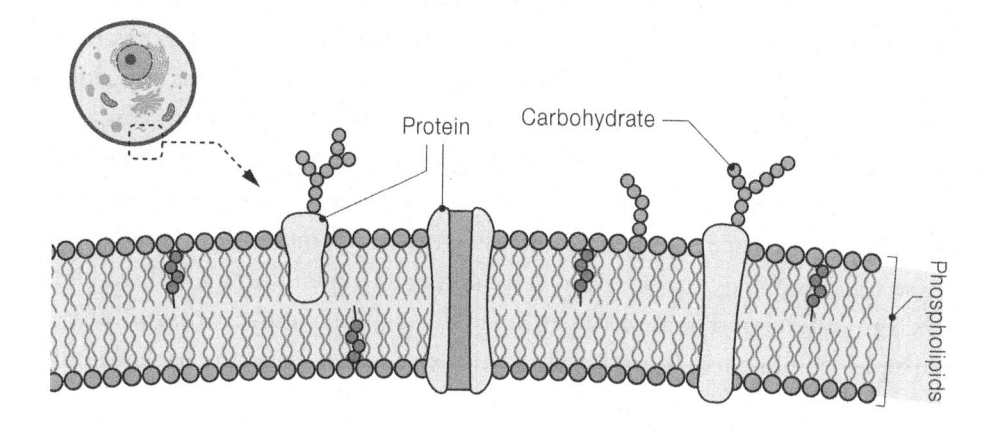

FIGURE 7

A schematic diagram of the cell membrane

Nucleus

The nucleus, which contains the entire genetic code of the cell, is usually located in the center of the cell (see Fig. 8). The diameter of a nucleus is only one-hundredth of a millimeter. However, the amazing "packaging technology" of nature ensures that the 1.8-meter-long (six-foot-long) DNA molecule fits in. In humans, the genetic material is not a single chain, instead, it is contained in twenty-three chromosome pairs, according to the chromosome number. Current knowledge states there are 20–25,000 genes in the chromosomes. A gene is a section of the DNA, encoding the amino acid sequence of a protein. Gene activation means that the cell has received a signal to start "producing" a specific protein (enzyme). When this happens, the cell creates a copy or copies of the specific DNA section. Such a copy is called a messenger RNA (mRNA). mRNAs carry information from inside the nucleus into the space between the nucleus and the cell membrane (the cytoplasm) to attach to the ribosomes, the "protein factories" of the cells. (The diagram also shows the pores on the nucleolus membrane ensuring the movement of molecules between the nucleus and the cytoplasm.) The ribosome is a large protein complex whose task is to produce an enzyme following a program encoded in the mRNA, which then can carry out its function in the cell. There is a constant flow of information between the cell membrane and the nucleus. As a result, the cell turns on and off genes that encode the proteins needed for its functioning. The cell also turns off genes whose protein products are currently not needed. Initiating and carrying out the cell division is a complex process that consists of a series of steps taking place within twenty to twenty-four hours. We know the role and function of hundreds of genes, proteins, and the regulatory molecules involved in these cell physiology processes. Yet, we still can't explain how a cell decides when to initiate and how to synchronize this complex process that involves multiple components. To understand this, we need to know the background of how mitochondria work.

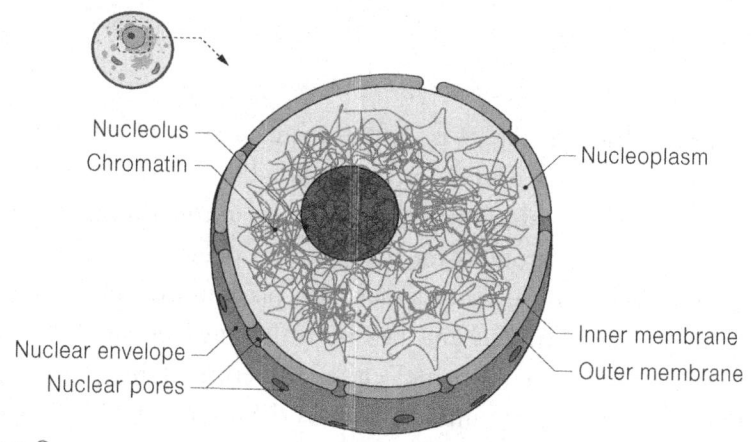

FIGURE 8

A schematic diagram of a cell nucleus

The mitochondrion

Mitochondria are the cell's powerhouses, found everywhere except in red blood cells. Certain kinds of cells have few mitochondria, while others, such as liver and/or muscle cells, have even hundreds of thousands of them. The mitochondrion's basic task is to store the energy released when "burning off" nutrients in adenosine triphosphate (ATP), universally accessible energy-storing molecules. ATP provides the energy for the above-mentioned protein synthesis. It is also indispensable for several other biochemical processes, such as the synthesis of DNA during cell division, when more than three billion purine and pyrimidine bases connect, doubling the cell membrane or contraction of muscle fibers.

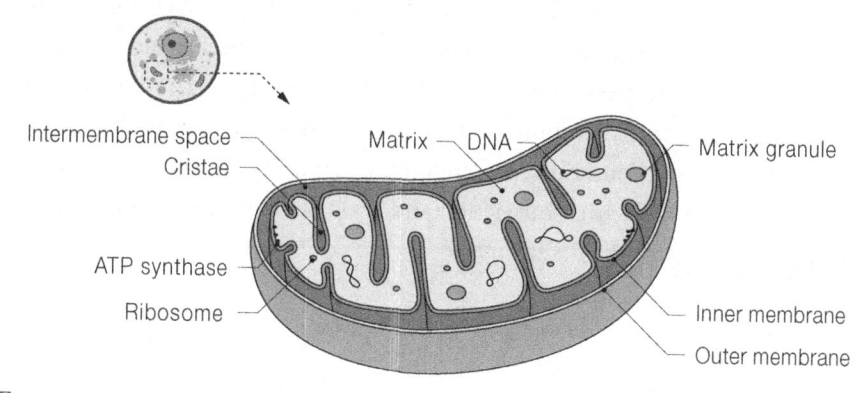

FIGURE 9

A schematic diagram of a mitochondrion

The mitochondrion in Fig. 9 is only one μm in size, which corresponds to the average size of a bacterium. Mitochondria always have an outer membrane, but also an additional membrane layer containing, along with other enzymes, ATPase proteins for the synthesis of ATP (see Fig. 10). The most important biochemical process that takes place in the mitochondrion is the oxidation of carbohydrates and fats from the nutrients, and the storing of the released energy in ATP molecules. In terms of biochemistry, the process that takes place in the mitochondria is the reverse of what happens in plants during photosynthesis. Plants are capable of using the energy of light to produce carbohydrates (sugars) from carbon dioxide in the air and water while releasing oxygen into the atmosphere. Mitochondria return the energy „captured" by the plants, producing carbon dioxide and metabolic water in the process. The hydrogen in the water molecule stems from the carbon atoms in carbohydrates and fats, and the oxygen comes from the atmosphere. (The origin of atmospheric oxygen can be traced back to the breakdown of the water molecules released in plants during photosynthesis, thus closing the cycle.) The ATP synthase responsible for the synthesis of ATP is a nano-motor capable of rotating around six to nine thousand times per minute. This nano-motor is driven by a stream of hydrogen. During biochemical processes, enzymes release hydrogen into the space between the inner and outer membranes, creating a significant difference in the concentration of hydrogen between the two sides of the inner membrane. This concentration difference provides the energy needed for the ATP synthase to rotate, the synthesis of ATP. It has been shown that the ATP synthase enzyme is capable of discriminating between two hydrogen isotopes, preferring the lighter isotope [45]. Two factors have a substantial influence on the concentration of deuterium in the metabolic water produced in the mitochondria: 1.) The concentration of deuterium present on the carbon atoms that are introduced into the mitochondria with nutrients, and 2.) The strength of discrimination during the activity of ATP synthase.

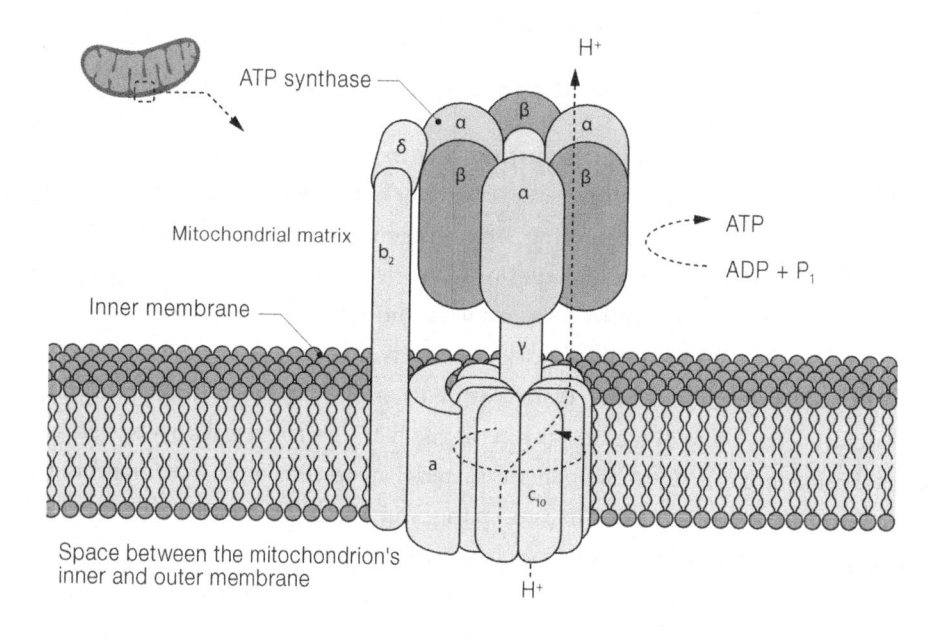

FIGURE 10
Subunits of the ATP synthase enzyme

One of the main goals of this brief introduction to cell biology is to show how processes in the cell can affect the deuterium/hydrogen ratio in two known cases. Hydrogen transport through the cell membrane may raise the deuterium/hydrogen ratio, while the mitochondrion is capable of lowering the same ratio. Variation of this ratio - whether it rises or falls - has a substantial effect on all enzymes and molecules in the cell, including the function of genes. Armed with this knowledge, you can now understand the results of the experiments conducted with DDW and what their implications are.

The submolecular regulatory system (SMRS)

This book does not aim to give a detailed overview of basic research from the past thirty years but to better understand the anti-cancer effects of DDW, conclusions regarding cell division are provided. As in the previous chapters of this book, references are provided to the most relevant publications.

The evolution of the regulatory mechanism

Experiments with heavy water confirmed that in primitive organisms, water can almost be entirely replaced with heavy water. For higher-level organisms, however, this is not possible without triggering a toxic, potentially even lethal effect [32, 36]. Studies examining the effect of deuterium-depleted water on primitive organisms concluded that low deuterium concentration did not affect the reproduction of bacteria [46]. A significant inhibitory effect was demonstrated, however, in the case of higher-level organisms, plants, animals, and in tissue cultures of animal or human cells. These observations hint at the fact that the regulatory systems of prokaryotes (organisms whose genetic material is not enclosed in a membrane-covered nucleus) are not sensitive to changes in the hydrogen/deuterium ratio. In the more complex eukaryotic organisms, a few ppm increase or decrease in the above ratio triggers a detectable change in the functioning of genes, enzymes, biochemical processes, or even in the entire body.

It is important to emphasize that this statement is only true for deuterium levels close to the natural ones. Heavy water exerts a substantial effect on prokaryotes and eukaryotes alike. The reason for this effect is the different chemical behavior of deuterium (its atomic weight is twice as much as that of regular hydrogen): the isotope effect. Nonetheless, the effects observed during deuterium depletion cannot be traced back solely to the isotope effect. Effects observed at increased deuterium concentrations were detectable only at differences on the order of a hundred or a thousand. Conversely, even a 1% difference in the concentration of deuterium was detectable in case of deuterium depletion.

One cannot just resolve this contradiction and explain the above disparity with a simple chemical reaction. It is obvious that in biological systems, even a small-scale change in the concentration of deuterium triggers complex, detectable and significant effects, involving a network of chemical reactions.

Imagine a network of chemical reactions as traffic in a city. When traffic lights are perfectly synchronized, traffic flows smoothly. If the cycle time of one or two green lights is shortened or extended even to a small extent, then possibly there won't be any immediately detectable change. A traffic jam may start to build up in one of the intersections. After some time, this congestion starts to affect drivers at other traffic lights, so they cannot cross the intersection. This is when the cross-traffic becomes blocked, eventually leading to the collapse of the city's traffic flow. All this can be traced back to an insignificant change in one of the elements of the system. Similarly, changes in the deuterium/hydrogen ratio influence the chemical reactions and the enzyme activity in biological systems through the above-mentioned kinetic isotope effect. A shifting deuterium/hydrogen ratio triggers changes in protein conformation and gene activity. These changes are detectable even at a 1% decrease in deuterium levels.

Evolution has created ever more complex biochemical mechanisms and physiological functions. Supposedly, a submolecular regulation was added to the one on the molecular level in the process. The submolecular regulation ensures that processes, despite their growing complexities, keep flowing in a regulated and synchronized manner. From an evolutionary perspective, it was a significant event when the ancestor of present-day mitochondria, a prokaryotic organism, was engulfed by a eukaryotic organism, the latter already possessing a nucleus with a double membrane. In eukaryotic cells, energy is produced by anaerobic respiration (without the presence of oxygen) in the cytoplasm, resulting in the production of two ATP molecules when breaking down a glucose molecule. Mitochondria break down glucose molecules only in the presence of oxygen. This provides thirty-six ATP molecules for the energy-intensive processes of the cell. Ancient eukaryotic cells that engulfed a mitochondrion enjoyed a selective advantage over other eukaryotic organisms. As a result, almost all eukaryotic cells have at least one mitochondrion, and cells that consume a lot of energy might have several thousand mitochondria. Despite the close cooperation and symbiosis between the two organisms, mitochondria enjoy a certain degree of autonomy. They have their own DNA, and their cell division is independent of their host cell's. This leap of evolution opened the possibility for the creation of a submolecular regulatory system (SMRS) based on the changes in the deuterium/hydrogen ratio.

The submolecular regulatory system (SMRS) and the biochemical processes

The difference between the atomic weight of hydrogen and deuterium is markedly present on the level of biochemical reactions. Consequently, a changing deuterium/hydrogen ratio exerts a substantial effect on biochemical reactions and the speed thereof. As hydrogen and deuterium are key constituents of every molecule, a change in the deuterium/hydrogen ratio influences the deuterium/

hydrogen ratio of every molecule, and ultimately, the behavior of every molecule in chemical reactions. The control over the molecular level is usurped by a submolecular mechanism (SMRS). This submolecular mechanism may be the key and requirement for the reliable, synchronized, and accurate functioning of biochemical, genetic, and physiological processes in advanced organisms. This also means that getting to know the submolecular regulatory system introduces a new therapeutic approach not just for the treatment of cancer, but for other metabolic or allergic diseases, or even in the field of sports medicine.

In vitro experiments involving DDW confirmed that the growth of tumor cells slowed down (or was inhibited) in a low-deuterium growth medium. This highlights the crucial role that naturally occurring deuterium plays in regulating cell division [12]. The first studies confirmed that increasing the deuterium concentration of the growth medium above 150 ppm (to 300–600 ppm) stimulates cell growth. The studies suggested that deuterium is indispensable for cell reproduction, as one of the signals of cell division is a temporary increase of the deuterium levels relative to the hydrogen levels (an increasing deuterium/hydrogen ratio) [12].

Even in the 1980s, years before our first publication on the inhibitory effect of DDW on cell division, an article was published about how the pH of cells shifts to alkalinity before cell division. The reason for this phenomenon is that when the growth hormone binds to the receptor, it stimulates the sodium–proton exchanger. Cells take up Na^+, dispose of H^+, resulting in an increasing pH within the cell. This was regarded as an inherent part of cell division [47, 48]. Perona and Serrano's findings, published in Nature [49], offered strong evidence for the crucial role of increasing pH before cell division. Perona and Serrano wrote that animal cells transformed with yeast ATPase gene exhibited a faster exchange of H^+ ions through the cell membrane. As a result of the elevating pH levels within the cell, the cells divide faster. Cell division was stimulated to such a degree that researchers had a hard time interpreting how the transformed cells exhibited a tumor phenotype *in vitro* and caused a tumor in nude mice [49].

Experiments with DDW and the pH increase before cell division suggest that the sodium–proton exchanger prefers the lighter hydrogen ion (proton) to the twice-as-heavy deuterium. This discrimination raises the deuterium/hydrogen ratio, which is the ultimate signal to start cell division [12, 47, 48, 49]. Another study confirms the increasing deuterium/hydrogen ratio as a result of H^+ exchange. It describes that the ATPase enzyme discriminates between the two isotopes of hydrogen. The enzyme does not accept deuterium and retains it in the cell [45]. Our research has confirmed that the sodium–proton exchanger is sensitive to changes in deuterium concentration. Following the cell's intended pH reduction, cells could restore the original pH much slower in a low-deuterium growth medium.

The high sensitivity of the cellular proton exchange to deuterium was confirmed in our experiments with the aquatic plant *Elodea canadensis* [17]. Transferring the plant from regular-deuterium water to low-deuterium water induced an immediate activation of the proton exchange system, shifting the cell's pH to alkalinity and that of the environment to acidity. To sum up the studies, the activation of cell division is primarily associated with the elevated deuterium/hydrogen ratio when activating the cell membrane's proton exchange system.

Szent-Györgyi, in his famous 1973 interview on Hungarian television, also cited an example from traffic to demonstrate cellular regulatory mechanisms. He stressed that traffic cannot be controlled exclusively using green lights. Red lights also have an important role, stopping traffic when necessary. In healthy cells, it is the mitochondria that keep cell division under control. Mitochondria are the powerhouses of the cell, oxidizing organic molecules as a carbon source and producing low-deuterium metabolic water to ensure a lower deuterium/hydrogen ratio in the cell. (The deuterium concentration of the resulting metabolic water depends on what percent of the carbon source originated from fats, carbohydrates, and amino acids.)

The two processes – one increasing the deuterium/hydrogen ratio and one reducing it – ensure the adequate frequency of cell division in all healthy tissues. This means that the newly formed cells make up for the cells lost due to programmed cell death (apoptosis), and therefore remain the same size of a given organ or tissue. A common trait of tumor cells is that their mitochondria are deficient. Consequently, in patients who ingest average-deuterium concentration water and a lot of carbohydrates, the deuterium/hydrogen ratio easily increases when the growth hormone binds to a receptor. An increased deuterium/hydrogen ratio enables cell division, as the mitochondria are unable to reduce deuterium levels in the cell. These changes cause cells to more frequently engage in cell division, and the cells get out of control. Applying deuterium depletion reduces deuterium concentration in every cell of the body, making up for the loss of the mitochondria's function and affecting other biochemical and genetic processes.

Otto Heinrich Warburg confirmed in the 1920s that a common characteristic of tumor cells is their altered metabolism as tumor cells use anaerobic oxidation to break down sugar [3, 50]. In the so-called Warburg effect, tumor cells do not oxidize sugar in the mitochondria, but through glycolysis in the cytoplasm. The "end product" is thus not exclusively water and carbon dioxide (as in the case of healthy cells), but partly lactic acid. In this respect, a common characteristic of tumor cells is that, despite the availability of oxygen, the Szent-Györgyi-Krebs cycle is barely (if at all) functioning, therefore no or not enough low-deuterium metabolic water is produced for the healthy functioning of the cell. In the mitochondria, the two carbon atoms and the bound hydrogen atoms transported by the acetyl-CoA may be derived from carbohydrates, fatty acids, or amino acids. The different deuterium concentration values of nutrients imply that the deuterium concentration of metabolic water produced by the mitochondria also depends on the proportion of the two carbon atoms of acetyl-CoA coming from the three nutrient categories. The more carbon atoms originate from fatty acids, the lower the deuterium concentration of metabolic water.

Another observation backing the metabolic theory of cancer is that a malfunctioning cell is no longer capable of producing low-deuterium metabolic water due to the impaired functioning of mitochondria to compensate for the higher deuterium/hydrogen ratio and elevated pH. This phenomenon facilitates the uncontrolled division of tumor cells.

The above explains the benefits of a ketogenic diet for cancer patients. In a ketogenic diet, 80 to 90% of the daily calorie intake is composed of fats, reducing the deuterium levels of metabolic water down to even 118 ppm.

The relationship between the submolecular regulatory system (SMRS) and genetic function

The genetics-based approach to the formation of cancer is backed by certain anomalies identified in the past decades. Currently, the mutations of several hundred genes have been found associated with tumor formation. Today's drug development focuses on a specific, erroneous gene and the protein encoded by it, and seeks to clarify its role in regulating cell division. Once the role of the erroneous gene and its product in the regulatory mechanism was clarified, researchers "only" have to design a molecule to correct the process.

The nature of this strategy implies that researchers attempt to intervene at numerous points in the complex networks, clinically evaluating hundreds of drug candidates each year. An overwhelming majority of these candidates fail in Phase II, as their anti-cancer efficacy cannot be confirmed, or there are severe life-threatening side effects. A *New York Times* article from March 23, 2013, provides a summary of the efficacy of 13 anti-cancer drugs approved by the FDA in 2012 and illustrates the failure of the anti-cancer drug development strategy. The data reveal that only one drug was capable of increasing the life expectancy of patients in clinical conditions by six months, and a further two by only four to six weeks.

During our studies, I also examined whether deuterium depletion influences the functioning of genes, and if so, how and to what extent.

Several experiments have demonstrated that the overexpression of COX-2 gene, and the resulting intense prostaglandin synthesis, is a common feature of precancerous cells and of any tumor cell. It is recognized as having a role in tumor growth and the formation of metastases [51]. It has also been shown that non-specific inhibition of COX-2 gene, for example with aspirin, or specific inhibition with celecoxib, reduces the risk of cancer [51]. At the beginning of our research, we also investigated the effect of DDW on COX-2 gene expression and prostaglandin synthesis. The results, which were first reported in the book "*Defeating Cancer!*" published in 2000 [52], showed that deuterium concentrations of 20 and 60 ppm inhibited, and 200 ppm stimulated the expression of COX-2 gene to a minimum extent. The degree of COX-2 gene inhibition correlated with prostaglandin synthesis in vitro, with a value of 242.9 ng/mL at a deuterium concentration of 150 ppm, decreasing to 76 ng/mL at 20 ppm. Simultaneous inhibition of cell division also occurred. In further gene expression studies on DNA microarrays, we observed changes in the function of hundreds of genes at different-than-natural deuterium concentrations. These initial findings demonstrated the decisive effect of deuterium on gene expression, and more importantly that unlike today's drug development strategy of targeting a specific gene, deuterium depletion can influence the entire genotype simultaneously.

Genetic research in the field of oncology usually follows gene function changes in mice exposed to a carcinogen(s). In our study, we used the chemical carcinogen 7,12-dimethylbenz(a)anthracene (DMBA), which activated several oncogenes after only 24 hours. In the experiments, the behavior of oncogenes (c-Myc, Ha-ras, p53) involved in the regulation of cell division and the development of cancer was investigated when the mice were given DDW as drinking water. These studies found that deuterium depletion inhibited or affected gene expression to different degrees in various organs (spleen, lung, lymphatic nodes, thymus, kidney, liver), but still to a significant extent in all organs. Later, in a long-term (one-year) liquid ingestion experiment, it was shown that DDW as drinking water also prevented tumor formation by inhibiting genes. Repeating the experiment with the K-ras, c-Myc, and Bcl2 genes provided similar results [54], clearly indicating that the concentration of deuterium in the cells has a significant influence on genes.

In subsequent experiments, we used nanostring technology [55] to monitor gene changes. This method is used to determine how many mRNA copies of a gene are made. Using this technology, we examined the expression of 536 kinase and 236 tumor-related genes in a lung tumor cell line (A459) kept in growth media with different concentrations of deuterium. Gene expression was compared to the 150 ppm control. That is, we examined the extent to which gene expression was affected by reduced deuterium levels of 40 and 80 ppm and increased deuterium levels of 300 ppm. The data confirmed the previous results, with the expression of hundreds of genes being significantly altered in the growth medium containing deuterium levels other than 150 ppm in this experimental system. Of the 536 kinase genes, 135 (25.3%) and 124 (52.5%) of the 236 tumor-related genes showed expression changes that warranted further analysis (gene expression changes were 30% greater than controls and the number of copies exceeded thirty).

The most intriguing question that emerged from the experiment was how a specific gene behaves in a lower and higher-than-natural deuterium concentration growth medium. Also, the question is whether there is a clear dose-dependent relationship between the inhibitory effect of deuterium depletion and the stimulating effect of increased deuterium levels, or whether the opposite is true (that is, if gene expression is stimulated by DDW vs. inhibited by an increased concentration of deuterium). A further possibility is whether different-than-normal deuterium levels, either elevated or reduced, affect gene expression in a single direction. Our results with the two sets of genes are presented in Tables 4 and 5 [56].

Reduced deuterium concentration (40–80 ppm)		Control (150 ppm)	Higher than natural deuterium concentration (300 ppm)	
Direction of change	Number of genes		Direction of change	Number of genes
0	0	0	0	0
–	1	0	0	0
–	0	0	–	0
0	0	0	–	0
+	0	0	0	0
+	0	0	+	0
0	0	0	+	134
+	0	0	–	0
–	0	0	+	0

„–": gene expression inhibited; „+": gene expression stimulated

TABLE 4

135 kinase gene expression changes as a function of deuterium concentration

The 135 kinase genes were classified into two groups based on their behavior. A reduced deuterium concentration inhibited the expression of one gene only, while an increased concentration stimulated the expression of 134 genes.

Reduced deuterium concentration (40–80 ppm)		Control (150 ppm)	Higher than natural deuterium concentration (300 ppm)	
Direction of change	Number of genes		Direction of change	Number of genes
0	0	0	0	0
–	0	0	0	0
–	0	0	–	0
0	0	0	–	0
+	0	0	0	0
+	1	0	+	1
0	0	0	+	118
+	0	0	–	0
–	5	0	+	5

„–": gene expression inhibited; „+": gene expression stimulated

TABLE 5

Expression changes of 124 tumor-related genes as a function of deuterium concentration

The behavior of 124 genes associated with cancer was classified into three groups based on nine possible combinations. In the case of a single gene, we found that both reduced and elevated deuterium concentrations increased the number of copies. For 5 genes, a reduced deuterium concentration inhibited it, while an elevated deuterium concentration stimulated gene expression. However, 95.2% of genes (118 genes) were stimulated by higher than natural deuterium concentration.

Combining the data of the two groups of genes (a total of 259 genes), it was found that a 300 ppm deuterium concentration stimulated gene expression by more than 30% in the case of 97.3% of the genes. The above helps to interpret the results of previous experiments with DDW.

The control of cell division in biological systems is not achieved by reducing deuterium concentrations below the natural levels, but by preventing deuterium concentration from exceeding threshold levels.

The submolecular regulatory mechanism (SMRS) controls the function of the entire genetic makeup through changes in the deuterium/hydrogen ratio. Metabolic processes in the cell, biochemical reactions, nutrition, and physical activity all influence changes in the deuterium/hydrogen ratio, transmitting signals to genes via the SMRS. Changes in gene function due to alterations in the deuterium/hydrogen ratio feed back to the SMRS-regulated process on the molecular level.

Gene expression analysis using nanostring technology indicates that almost 100% of genes respond to higher deuterium concentration and activated in that range. This approach helps to interpret why the sodium–proton exchanger system is capable of inducing cell division by increasing the deuterium/hydrogen ratio. Tumor cells, due to malfunctioning mitochondria, are unable to perform their natural deuterium-reducing function and produce low-deuterium metabolites. This prevents or delays the creation of a high deuterium/hydrogen ratio needed to initiate cell division. The production of low-deuterium metabolic water is a "product" of healthy cellular metabolism, so deuterium depletion is a natural biochemical process that controls the cell's entry into the division phase.

The results from using water at elevated deuterium levels of 300 ppm, which demonstrated a significant increase in the expression of cancer-related genes, resolve the apparent contradiction between the genetic or metabolic approach to cancer and brings the two approaches to a common platform.

Research of the genetic causes of cancer has shown that in 15–20% of breast tumor cases, there is evidence of overexpression of HER2 gene, which encodes a human epidermal growth factor receptor. Overexpression of this gene means that 40–100 times more growth factor receptors are present on the cell surface, increasing the number of stimulating signals and leading to uncontrolled cell division. The drug trastuzumab (Herceptin), a monoclonal antibody that binds to the HER2 membrane protein in the G1 phase, can stop the cell cycle, blocking known signaling pathways and exerting a significant therapeutic effect.

A drug named Gefitinib has a similar efficacy effect. Gefitinib works by blocking one of the epidermal growth factor receptors (EGFR). The drug was first used in patients with lung cancer, but it only has a therapeutic effect in patients with a confirmed overexpression of the EGFR gene.

It is hypothesized that the overactivation of the HER2 and EGFR genes leads to a significant increase in cytoplasmic pH through the activation of the sodium-proton exchanger system and an increase in the deuterium/hydrogen ratio in the cell, which is ultimately responsible for the initiation of cell division. The two drugs mentioned above may also prove to be so effective because, in addition to blocking receptors, they put on the brakes before the cell has a chance to start dividing through the stimulation of the sodium-proton exchanger system. This also explains the low efficacy of many targeted therapies, as the signaling pathways identified and the inhibitors developed do not have a substantial effect on processes that are determined at another level of the regulatory system on the submolecular level, rather than the molecular level.

The results of our gene expression studies suggest that structural rearrangements of genes do not necessarily have to occur, as in the case of HER2 and EGFR receptors. Instead, the overexpression of genes may also occur when deuterium levels within the cell exceed the physiologically optimal level, which can be triggered by metabolic processes. The reverse of this statement is also true: deuterium depletion can counteract the processes that stimulate cell division at the molecular level, triggered by genetic differences.

A correct interpretation of the submolecular regulatory mechanism (SMRS) based on the deuterium-hydrogen isotope pair allows both approaches (the genetic and metabolic) to be plausible since whether we explain the development of cancer based on genetic defects or cellular metabolism, both approaches in fact can be traced back to perturbations in the SMRS.

The important physiological role of the submolecular mechanism and the associated deuterium-hydrogen isotope pairs is confirmed by scientific results published in recent years. These results show that altering deuterium levels reduces susceptibility to depression [57], improves long-term memory [58], and deuterium depletion has also been shown to have radioprotective [25] and anti-aging [59] effects.

It is reasonable to assume that research will confirm the crucial role of the deuterium/hydrogen ratio in many other biochemical and genetic processes.

The submolecular regulatory mechanism and tumor necrosis

Besides the existence of SMRS, another question is what sort of mechanism is responsible for the anti-tumor effect of DDW. To break this question down, we should ask: Which process induces the necrosis of tumor cells, the reduction in tumor size, and causes complete tumor regression, as well as why does tumor regression not always occur?

Minimum or no improvement was seen in some patients receiving deuterium depletion. During the consultations, we were able to identify the causes and commonalities that could be linked to the partial or total ineffectiveness of deuterium depletion. (See: *Other complementary therapies decreasing the efficacy of deuterium depletion*) One of the essential factors was the use of antioxidant vitamins.

It is widely accepted that cancer can be prevented by taking antioxidants regularly. Antioxidants can reduce the number of free radicals produced in cells, thereby protecting the integrity of the cell's genetic material, the DNA, so that fewer genetic defects are produced and so the likelihood of cancer is reduced. The evidence shows, however, that despite detailed analyses of hundreds of thousands of people taking antioxidant vitamins, it could not be confirmed that taking these vitamins significantly reduces the likelihood of developing cancer. A recently published paper [60] also investigated the effects of high antioxidant preparations on cancer incidence and mortality. The paper concludes that caution should be exercised when using high doses of antioxidant preparations. Beta-carotene in high doses, for example, increases the incidence of cancer in smokers and also increases cancer mortality. They conclude by stressing that antioxidants, in particular beta-carotene and vitamin E, do not prevent cancer. On the contrary, beta-carotene increases the risk of smoking-related cancer types (lung, head and neck, upper gastrointestinal tract, and bladder cancer). For the above reasons, in line with the opinion of other specialists and oncologists, we also advocate the consumption of vitamins in their natural form, alongside a balanced, healthy diet.

With the recognition of the relationship between antioxidant intake and reduced DDW efficacy, it was hypothesized that a decrease in deuterium concentration would increase the concentration of free radicals, inducing programmed cell death (apoptosis). However, this would be prevented if cells had an excess of antioxidant vitamins capable of neutralizing the radicals generated by deuterium depletion. This has been confirmed by independent studies, and the results of a team led by Professor Roman Zubarev at the Karolinska Institutet [61] should be mentioned. In proteomic studies, a full qualitative and functional analysis of the proteome of a DDW-treated tumor cell line found that deuterium depletion induced oxidative stress, in which mitochondrial proteins played a predominant role. Oxidative stress was induced by an increased flow of protons (H+) from the mitochondrial matrix to the intermembrane space, which ultimately induced cell apoptosis. However, proteomic studies have also indicated that oxidative stress triggers a feedback mechanism in the cell to eliminate such stress. Several previous studies have confirmed an increased activity of the enzyme superoxide dismutase (SOD) in response to deuterium depletion, which has a primary role in neutralizing reactive oxidative radicals. This also hints at DDW's reactive radical generating effect and indicates the accuracy and speed of the mechanism that has evolved to deal with oxidative stress [59, 62]. This confirms that free radicals arising from deuterium depletion are primarily responsible for cell death. This explains why deuterium depletion was found to be less effective or even ineffective in patients taking high doses of antioxidants, and why, in particular when tumor masses were significant, a given deuterium concentration was found to be less effective.

A key to the successful application of deuterium depletion is to determine the time interval within which a given deuterium concentration (DdU) can effectively induce cell death in the deuterium-sensitive phase of the cell cycle. Changing the deuterium levels occurs before the cell triggers the mechanisms that neutralize oxidative stress induced by DDW. Further tumor necrosis can be achieved by optimally timing the DdU increase, taking advantage of the effect of oxidative stress induced by DDW, and by blocking the mechanisms that counteract it.

The following figures illustrate the major elements of how the submolecular regulatory mechanism works, the common features of healthy and tumor cells, and the different characteristics of the two cell types.

A resting healthy cell in a normal deuterium-containing growth medium

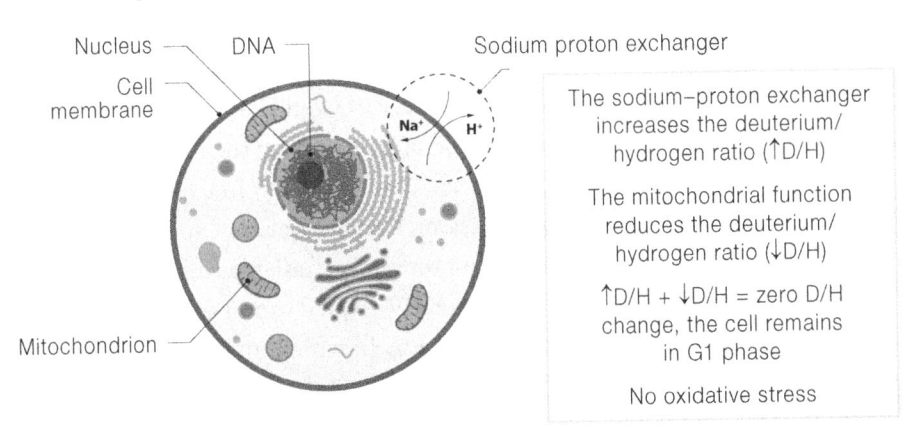

The sodium–proton exchanger increases the deuterium/hydrogen ratio ($\uparrow$D/H)

The mitochondrial function reduces the deuterium/hydrogen ratio ($\downarrow$D/H)

$\uparrow$D/H + $\downarrow$D/H = zero D/H change, the cell remains in G1 phase

No oxidative stress

A healthy cell in normal deuterium-containing growth medium, when cell division is induced

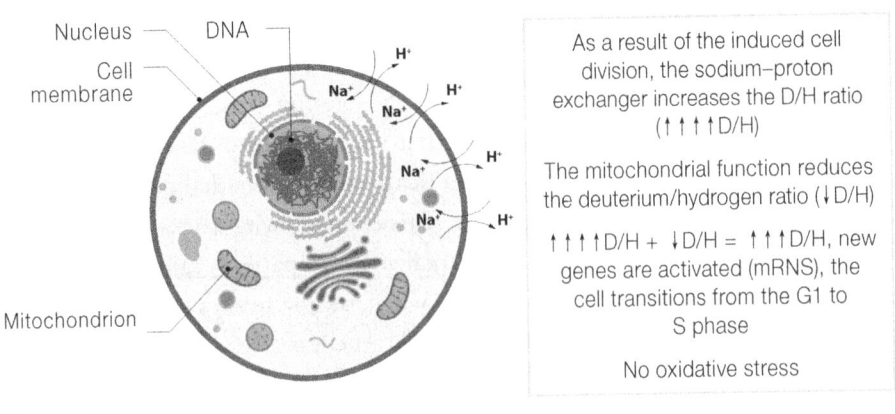

As a result of the induced cell division, the sodium–proton exchanger increases the D/H ratio ($\uparrow\uparrow\uparrow\uparrow$D/H)

The mitochondrial function reduces the deuterium/hydrogen ratio ($\downarrow$D/H)

$\uparrow\uparrow\uparrow\uparrow$D/H + $\downarrow$D/H = $\uparrow\uparrow\uparrow$D/H, new genes are activated (mRNS), the cell transitions from the G1 to S phase

No oxidative stress

FIGURE 11
Processes in resting healthy cells (G1 phase), and after the induction of cell division

In resting healthy cells, the sodium–proton exchanger system (capable of increasing the deuterium/hydrogen ratio [↑D/H]) and the mitochondria (producing low-deuterium metabolic water during the cellular respiration [↓D/H]), operate simultaneously. The two processes are in equilibrium, so the deuterium/hydrogen ratio does not change significantly, metabolic processes take place as normal (Fig. 11), and the cell remains in G1 phase. If the cell is exposed to a growth stimulus (e.g., a growth hormone binds to the receptor), the sodium–proton exchanger system is activated and the deuterium/hydrogen ratio starts to increase in the cytoplasm, which cannot be compensated by mitochondrial processes. As a result, new genes are activated in the nucleus, their transcribed mRNAs appear in the cytoplasm, and, in accordance with the induction, the proteins they encode are synthesized. This initiates cell division, and the cell, while maintaining a regulated metabolism, leaves the G1 phase, and starts transcribing DNA (S phase). Cell division then takes place in twenty to twenty-four hours.

The tumor cell in normal deuterium-containing growth medium, before the induction of cell division

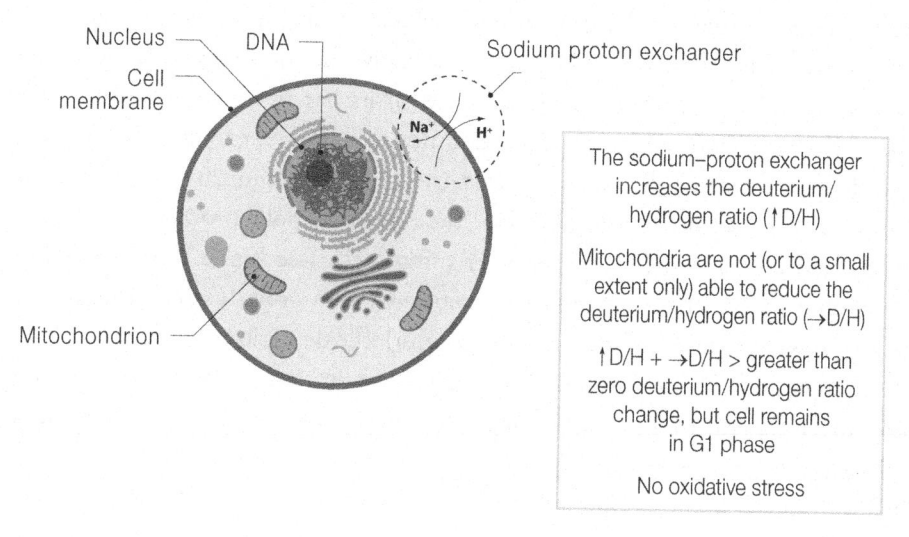

Labels: Nucleus, DNA, Sodium proton exchanger, Cell membrane, Na+, H+, Mitochondrion

The sodium–proton exchanger increases the deuterium/hydrogen ratio (↑D/H)

Mitochondria are not (or to a small extent only) able to reduce the deuterium/hydrogen ratio (→D/H)

↑D/H + →D/H > greater than zero deuterium/hydrogen ratio change, but cell remains in G1 phase

No oxidative stress

The tumor cell in normal deuterium-containing growth medium, after the induction of cell division

[Diagram labels: Nucleus, DNA, Cell membrane, Mitochondrion, with H⁺ and Na⁺ ions shown]

> The sodium–proton exchanger increases the deuterium/hydrogen ratio (↑↑↑↑D/H)
>
> Mitochondria are not (or to a small extent only) able to reduce the deuterium/hydrogen ratio (→D/H)
>
> ↑↑↑↑D/H + →D/H = ↑↑↑↑D/H, new genes are activated (mRNA), cell transitions from G1 to S phase
>
> No oxidative stress

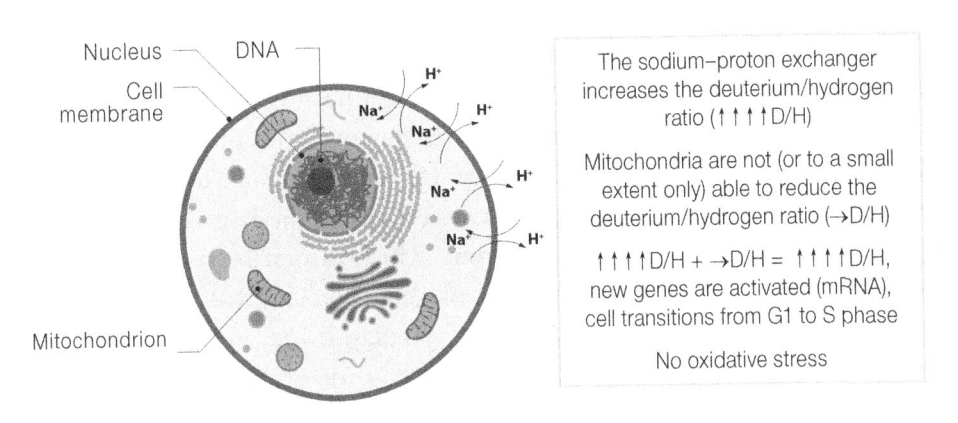

FIGURE 12

Processes in tumor cells during the resting phase (G1 phase) and after the induction of cell division

The sodium–proton exchanger system, which is able to increase the ratio of deuterium to hydrogen (↑D/H), is also active in the membrane of resting tumor cells. The mitochondria, however, with its barely functioning (if at all) Szent-Györgyi-Krebs cycle, are not able to (or to a limited extent only) produce low-deuterium metabolic water, so that the cell's deuterium/hydrogen ratio is shifted in favor of deuterium (↑D/H). The cell's metabolic processes take place in a balanced way, and as is typical of tumor cells (see Fig. 12), with the cell remaining in G1 phase. If the cell is subjected to a growth-stimulating effect (e.g. a growth hormone binding to the receptor), the deuterium/hydrogen ratio in the cytoplasm is further increased through the activation of the sodium–proton exchanger system, which can no longer be compensated by the malfunctioning mitochondria. As a result, new genes are activated in the nucleus, their transcribed mRNAs appear in the cytoplasm and, in response to the induction, their encoded proteins are synthesized, initiating cell division. The cell, while maintaining a regulated metabolism, leaves the G1 phase and starts transcribing DNA (S phase). The cell division takes place in twenty to twenty-four hours. There are no fundamental differences in cell division between healthy and tumor cells.

The key difference between healthy and malfunctioning cells is that while healthy cells are less likely to engage in cell division due to proper mitochondrial function maintaining a lower deuterium/hydrogen ratio, tumor cells' deuterium/hydrogen ratio is higher, even in a resting state, due to impaired mitochondrial function. Therefore, the deuterium/hydrogen ratio is more likely to rise to the level required to initiate cell division in response to growth stimulation.

The healthy cellular response to deuterium depletion

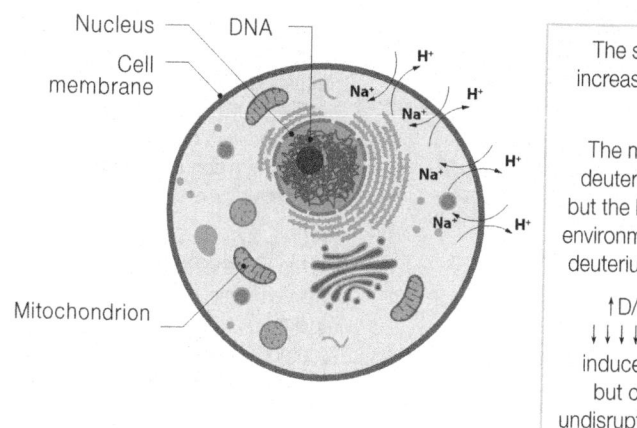

The sodium–proton exchanger increases the deuterium/hydrogen ratio (↑↑↑↑D/H)

The mitochondrion reduces the deuterium/hydrogen ratio (↓D/H), but the lower D concentration of the environment significantly reduces the deuterium/hydrogen ratio (↓↓↓D/H)

↑D/H + ↓D/H + ↓↓↓D/H = ↓↓↓↓D/H, deuterium depletion induces oxidative stress in cells, but cells adapt quickly due to undisrupted metabolism and remains in the G1 phase. Cell division is induced only after adaptation to oxidative stress

FIGURE 13

Processes in healthy cells when deuterium depletion is applied, after the induction of cell division

A change in the deuterium concentration of the environment causes an immediate and significant decrease in the deuterium/hydrogen ratio in cells, to which the cells respond immediately. Basic research [17] and human Phase II clinical results [44, 63] also suggest that lower deuterium concentrations activate membrane transport processes that allow cells to restore the original deuterium/hydrogen ratio by selectively preferring the lighter isotope during transport processes. Another simultaneous process is the appearance of reactive oxygen radicals: oxidative stress [59, 61]. Examination of the above parameters in healthy cells clearly shows that cells adapt rapidly to these changes and eliminate free radicals by activating the cell's redox processes. Cell metabolism is normalized, and in the case of induction, cells are able to initiate cell division.

Response of the tumor cell to deuterium depletion

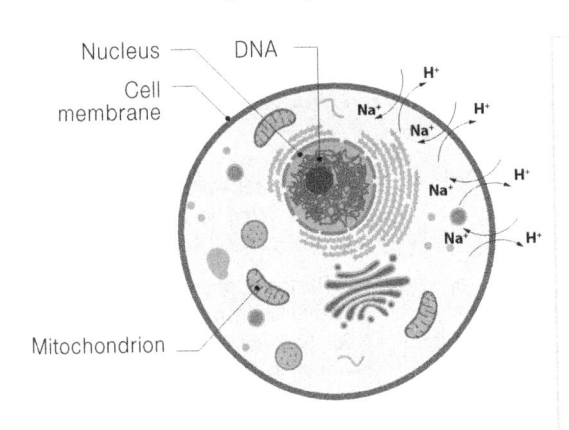

The sodium–proton exchanger increases the deuterium/hydrogen ratio (↑↑↑↑D/H)

Mitochondria are not (or only to a small extent) able to reduce the deuterium/hydrogen ratio (→D/H), but the deuterium/hydrogen ratio is still reduced due to the lower deuterium concentration in the environment ↓↓↓D/H)

↑D/H + →D/H + ↓↓↓D/H = ↓↓D/H in the cell, the reverse processes induce oxidative stress, which the impaired metabolic processes of the tumor cells cannot adequately respond to, the decrease in the D/H ratio induces apoptosis, the tumor cell necrotizes. When cell division is induced, the opposite messages of active membrane transport processes, deficient metabolic processes and a decreasing deuterium concentration interfere with the restoration of the cell's metabolism, stimulating the necrosis of the tumor cell

FIGURE 14
Processes in tumor cells when deuterium depletion is applied, after the induction of cell division

Changes in the deuterium concentration in the environment also cause an immediate and significant decrease in the deuterium/hydrogen ratio in tumor cells. In *in vitro* experiments investigating the sensitivity of cells to deuterium concentration decreases, we found that a concentration decrease of as little as 1 ppm every eight hours resulted in slower growth rates [64], and a decrease of 5 ppm twice within eight hours completely stopped cell division. These *in vitro* results demonstrate that tumor cells are not able to (or only much more slowly) respond to a decrease in deuterium concentration in the same way as healthy cells, and thus are not able to adapt to the new situation. A further drop in the deuterium concentration at a later time, but before any adaptation occurs, results in the inability of the diseased cell to cope with the oxidative stress and induces the cell's self-destructive process, apoptosis. Results indicate that the tumor cells and tissues that are most susceptible to deuterium depletion are those in the most active growth phase.

The recommendations for the use of DDW and its optimal fitting to conventional treatments presented in the rest of the book are developed with these considerations in mind. Experience has shown that several circumstances and parameters influence the outstanding efficacy of deuterium depletion. These circumstances and parameters, just like in case of the conventional therapies, may impair the effectiveness of the procedure.

CHAPTER THREE

Clinical Results of DDW Application in Cancer Patients

The concept of DDW dosage

The concept of DDW dosage, similarly to the dosage of other drugs, has two different meanings: it defines the daily dose of DDW (the daily intake), and the deuterium concentration of DDW (the strength of the active substance).

The daily dose is the amount of DDW that covers a patient's daily fluid intake. It is important to note that DDW should ideally cover 75–80% of the daily fluid intake. Liquids with normal deuterium levels in larger amounts may significantly weaken the effect of deuterium depletion (the effect of the active substance). For optimal deuterium depletion, it is advisable that patients ingest DDW only, and cover their additional fluid intake needs (on top of 75–80% of their daily intake) by consuming fruits, vegetables, and other nutrients.

When ingesting one unit (a specific volume) of DDW, the greater the deuterium level reduction in the system is:

  a) The lower the deuterium concentration of DDW,

  b) the greater the difference in concentration between DDW and the body's deuterium level,

  c) the smaller the body's volume of water that the daily DDW intake mixes with (in other words, the smaller the body mass is).

To ensure a successful treatment, the following key aspects should be observed:

  a) Patients should be administered the proper concentration of deuterium,

  b) the daily DDW volume should reach the required volume,

  c) in addition, it is necessary to limit the fluid intake with normal deuterium levels.

In two cases deuterium levels are not reduced to produce the desired effect. The first case occurs when patients ingest water with deuterium concentrations higher than required, or with less volume than required (upon their body mass). The second case happens when patients consume a large amount of other liquids with normal deuterium concentration other than DDW (tap water, bottled mineral waters, soft drinks, fruit juices, milk, etc.).

The following formula was designed for a better comparison of the varying parameters and to evaluate the data: $\frac{145}{72} \times 1 = 2.01$

$$\text{Dosage (DdU)} = \frac{150 \text{ (ppm)} - (\text{DDW-concentration (ppm)})}{\text{body weight (kgs)}} \times \text{DDW-volume} \left(\frac{\text{liter}}{\text{day}}\right)$$

DdU increases with a decreasing concentration of deuterium and increasing daily intake of DDW. The DdU, however, is reciprocally proportional to body weight [65]. Two examples are provided to illustrate the variation in DdU:

(a) If a person weighing 60 kilograms consumes 1.66 liters of water with a deuterium concentration of 65 ppm a day, then the DdU is 2.36.

(b) If a person weighing 80 kilograms consumes 1.33 liters of DDW per day at a concentration of 105 ppm, the DdU is 0.75.

One cannot give a single number to answer the question of what a proper DdU is. As discussed later, a number of other parameters (tumor type, stage, type and effectiveness of the ongoing conventional treatment, etc.) have to be considered to establish the ideal dosage. Discussed below are the key aspects that play a role in determining the optimal dosage of DDW. For more information on the consumption of DDW, see the chapter, *"General advice on the application of deuterium depletion."*

PROSPECTIVE PHASE II CLINICAL TRIAL ON PATIENTS WITH PROSTATE TUMORS

Data from a Phase II, double-blind clinical trial involving forty-four prostate cancer patients were evaluated [66]. The distribution of patients in the two groups (active substance vs. placebo) was twenty-two to twenty-two. During the clinical trial, all patients consumed the blinded preparation for four months. Patients in the treatment group received DDW with a concentration of 85 ppm, whereas patients in the control group consumed water with a concentration of 145 ppm. There was no difference in the distribution of patients by stage and the type of conventional treatment in the two groups. The study aimed to verify the anti-cancer effect of DDW. During the four months of the study, the following parameters were evaluated each month: PSA level, determining the prostate size with rectal ultrasound examination, urination problems, uroflowmetry, blood count, liver function test, kidney function test, etc. To assess efficacy, the internationally accepted naming convention was used to evaluate the data (CR: Complete Response, no detectable tumor, PR: Partial Response, regression exceeding 50%, PD: Progression of Disease, increase of tumor exceeding 25%, NC: No Change, patients do not fall into either the PR or the PD category). A summary of the results is given in Table 6, showing that the number of patients responding well to treatment (PR) was significantly higher in the treatment group.

Efficacy	Treatment group	Control group	Number of cases
PR	7	1	8
NC	11	13	24
PD	4	8	12
Total case number	22	22	44
	Armitage Exact Test	Fisher Exact Test	
p	0.027	0.046	

TABLE 6
Distribution of efficacy of Phase II, double-blind clinical trial on prostate tumor patients in the treatment (DDW) and control (normal water) groups

Comparing the cumulative PSA values, an 80% decrease was reported in the treatment group. The PSA decreased from 406 ng/mL at the initial examination to 80 ng/mL at the sixth follow-up medical examination. In the control group, which received hormone therapy only, there was a 47% reduction in

PSA, from 521 ng/mL to 277 ng/mL. Similarly, there was a significant change in cumulative prostate volume, which showed a decrease of 160 cm^3 in the treatment group compared to only 54 cm^3 in the control group (p = 0.0019). The one-year survival figures confirm the anti-cancer effect of DDW. Despite only four months of the clinical trial, "only" two of the twenty-two patients in the treated group died, compared to nine of the twenty-two patients in the placebo group (p = 0.029) [66].

In addition to demonstrating the effect of DDW on prostate tumors, the clinical trial confirmed that the active substance is safe to use, with no signs of toxicity or adverse side effects observed.

Follow-up (retrospective) studies

Deuterium-depleted water became available in larger quantities in 1992 and has been commercially available in Hungary since the fall of 1994. DDW with a deuterium concentration of 25 ppm was registered as a veterinary prescription drug in 1999 by the competent authority under the name Vetera-DDW-25 for the complementary treatment of pets with cancer. In the same year, the deuterium-depleted drinking water products Preventa-105 and Preventa-85 were approved, which allowed wider access to DDW. Since the early 2000s, new Preventa products have been introduced (Preventa-125, Preventa-65, Preventa-45, Preventa-25), further widening the range of DDW concentrations available.

Three major factors are supporting the significance of observations, findings, and feedback from people consuming DDW:

(a) The evaluation of a relatively large population of about 2,000 subjects who consumed DDW for a minimum of three months,

b) the first experience goes back twenty-seven years (1993), thus about half of the population has had a follow-up period of two to twenty years,

(c) the patient population is representative of the diversity of cancers in terms of tumor types, stage and treatment protocols.

In recent years, we have evaluated data from patients with mammary, prostate, pancreatic, and lung cancers and published the results. The results are presented below.

Breast cancer

In 2012, data from 232 breast cancer patients were processed and the results were published in 2013 [65].

The average patient age was 51 years (median: 50 years), the average time from the diagnosis to the start of DDW consumption was 3 years (median: 1.2 years), the average duration of DDW consumption was 2 years (median: 1.15 years), the duration of follow-up since the diagnosis was 5.8 years (median: 4.1 years), and the average duration of the follow-up since the start of DDW consumption was 2.8 years (median: 1.5 years). While 129 patients started consuming DDW in stage IV (advanced stage), at the time of the diagnosis, only 74 patients were classified as stage IV according to the diagnostic criteria. 55 patients started the treatment in an early stage, while 48 patients were in complete remission and tumor-free after successful treatments. The large patient population offered a possibility to consider various aspects to calculate the median survival time (MST) in homogeneous patient groups based on the Kaplan–Meier curve. MST was calculated compared to two dates: the date of the diagnosis and the start of DDW consumption. (Median survival is defined as the timespan within which half of the patients are still alive.) The cumulative follow-up period for 232 patients was 1,346 years.

The MST starting at the date of the diagnosis was 18.1 years for the patients diagnosed at an early stage (n = 158). For the patients diagnosed in stage IV (n = 74), the MST was 4.3 years, approximately twice those stage IV patients receiving conventional treatment only.

In a further grouping, it was investigated how the time elapsed between the diagnosis and the start of DDW consumption affected the life expectancy of the patients. 114 patients started DDW within one year from the diagnosis. In this group, due to the low mortality rate (cumulative follow-up time 396 years/22 deaths), no MST could be calculated. For patients (n = 118) who started DDW later than one year, MST was calculated as 4.1 years.

Of the 232 patients, 53 have had regular repeat courses over the years, 179 have consumed DDW only once for a shorter or longer period. MST was 24.4 years in the first case and 9 years in the second case.

The most important result in terms of survival was that of the 48 patients who started DDW in a tumor-free state, only one patient died during the cumulative follow-up period of 221 years. This in itself is strong evidence that the integration of DDW in follow-up treatments reduces the likelihood of the disease recurrence to a fraction of the initial figures.

The 74 patients who were already classified as stage IV at the time of the diagnosis were evaluated separately. A total of 135 distant metastases were confirmed in this group. In their cases, a demonstrable relationship was found between the dosage and efficacy of DDW. For each patient, the DdU value was calculated separately each time there was a change in DDW concentration, daily amount consumed, or patient weight. Simultaneously, we recorded the changes in the patients' follow-up examinations. Four categories were established: complete regression (CR), partial regression (PR), no change (NC), and the progression of disease (PD). The mean DdU for patients in the CR group was 1.68 and 1.28 for the patients in the PR group. Conversely, patients in the NC and PD groups had a DdU of less than 1 (NC: 0.66, PD: 0.92).

When splitting the patients into two groups based on the DdU (DdU smaller than 1, equal to 1 or greater than 1) as shown in Table 7, a 70% CR or PR was achieved in the group with a DdU greater than 1. Conversely, 69% of the patients in the other group belonged to the NC or PD category (p = 0.0028).

DdU	CR	PR	NC	PD
> 1	12 (30%)	16 (40%)	3 (7%)	9 (23%)
≤ 1 p = 0.0028	3 (7%)	10 (24%)	10 (24%)	19 (45%)

TABLE 7

Efficacy curve as a function of DdU of the DDW treatment in 74 Stage IV patients

The fact that in exceptional cases, we detected a CR even if the patient's DdU was below 1 shows the complexity of the situation. It also occurred that the disease progressed despite a DdU greater than 1. The general conclusion is that the recommended DdU for DDW treatment is approx. 1 with minimal variation. A detailed analysis of the results in breast cancer patients highlights the most important contexts and considerations. It is recommended to add deuterium depletion to the treatment as soon as possible after a diagnosis so that the DdU exceeds 1. Applying the procedure is advisable even if a patient is already tumor-free. Patients in remission are recommended to repeat DDW courses over several years (see: *Advice on establishing the dosage*).

Prostate tumor

In addition to the patients from the previously described prospective Phase II clinical trial, we evaluated data from 91 additional prostate cancer patients [66]. Of these patients, 45 had tumors in the prostate only, while 46 patients had already developed distant metastases. Of the 46 patients, 32 had bone metastases only, and 14 had metastasis in other organs, too. Of the patients treated with bone metastasis, 20 patients developed the metastases within one year after the diagnosis, while in 12 cases, metastases developed within more than a year. MST was calculated first for 91 patients, without forming small homogeneous groups of patients. The MST was calculated to be more than 11 years. The long MST was explained by the fact that the median survival could not be determined for the 45 patients without metastases due to low mortality rates. In this case, there were "only" four deaths during the 157 years of the cumulative follow-up period. The MST of patients (n = 20) treated with bone metastasis within one year was 64.8 months, three times longer than the MST according to the historical control. For the 12 patients who developed bone metastases later than one year after diagnosis, it was not possible to determine MST either, due to "only" two deaths during the cumulative follow-up period of 103 years. These MST values, and the low mortality rate observed despite the late stage, further confirm the results of the previously conducted prospective Phase II clinical trials.

Monitoring PSA levels in case of a prostate tumor offers a rare opportunity to verify the efficacy of deuterium depletion. Some of these cases are presented in the chapter, "*A demonstration of the effectiveness of deuterium depletion through case studies.*"

Pancreatic tumor

Pancreatic cancer is one of the most aggressive cancers with one of the worst prognoses. When my previous book *Defeating Cancer!* was published back in 1999, we had a few years of experience and no successful cases to report. This also indicated that there might be significant differences in sensitivity between tumor types. The options were further limited by the fact that only 90–100 ppm DDW was available at that time, but as lower-deuterium water became available, survival results improved. At the AACR (American Association for Cancer Research) conference in San Diego in 2014, data from thirty-two patients with pancreatic cancer who consumed DDW in addition to conventional treatments and from thirty patients who received only conventional treatments were compared [67]. The MST for patients in the control group was 6.36 months, which is consistent with the results of other independent clinical studies on pancreatic cancer. Because of the short life expectancy, two groups were formed based on whether the patients started consuming DDW within sixty days of diagnosis or later. The MST was thirty-nine months for the first group (n = 18) and sixteen months for the second group (n = 14). Despite an extension of life expectancy in both groups as a result of DDW treatment, results clearly showed that substantial research has to be done to increase the efficacy of DDW in this type of cancer.

Lung tumor

We first reported four cases where brain metastases were confirmed before the start of DDW consumption. Pathologists identified non-small cell lung cancer (NSCLC) in three of the four patients and small cell lung cancer (SCLC) in the last case. The SCLC patient had a brain metastasis measuring 20×30×40 mm, and the NSCLC patients had multiple brain metastases, four and two, respectively [68]. Sources suggest that the median survival time for untreated patients at this stage is one to three months, which is minimally extended by radiotherapy and steroids following protocols.

The first patient with NSCLC (a forty-five-year-old male patient) underwent immediate surgery for brain metastases, with the largest tumor removed but metastases up to 5 mm in diameter not removed. Subsequently, it was revealed that the patient's primary tumor was in the lung (60 × 40 mm). The patient had started consuming DDW immediately after diagnosis in parallel with treatment with Carboplatin-Etoposide. Four months later, complete regression of the small brain metastases and significant regression of the primary lung tumor was observed (the tumor size shrunk from 60 mm in diameter to 50 mm). The patient consumed DDW for 17.5 months while enjoying a good quality of life. However, the patient later died more than two years after diagnosis and nine months after discontinuation of DDW consumption.

A forty-one-year-old male patient, also diagnosed with NSCLC, had one of four brain metastases surgically removed, followed by cranial irradiation and chemotherapy. However, the disease progressed despite the treatment, with a life expectancy of only a few weeks. The patient then started consuming DDW, which halted the progression of brain metastases, and a few months later, an X-ray of the lungs showed regression. The patient's quality of life was satisfactory despite his advanced stage of the disease. We followed the development of his condition for ten months afterward, with no further information later on.

The third NSCLC patient (a sixty-one-year-old female) developed two brain metastases after four symptom-free years following removal of the primary lung tumor after which she started to consume DDW. With a combination of radiotherapy, chemotherapy, surgery, and deuterium depletion, the patient died 33.4 months after the appearance of brain metastases.

The fifty-four-year-old female patient (SCLC) developed a brain metastasis originating from a 30 × 50 × 32 mm primary lung tumor. The patient started consuming DDW in addition to radiotherapy and chemotherapy. A cranial MRI scan described significant regression (which continued later) after the first three months of DDW treatment. Scans were taken almost two years later and then forty-three months later showed only structural changes due to brain metastasis (Fig. 15). At the time of publication of the cited article, the patient was still symptom-free seven years after the diagnosis and died twelve years after the discovery of lung cancer, in 2013. The case of this patient [68] is not unique among lung cancer patients consuming DDW.

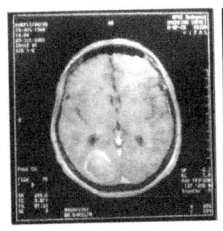

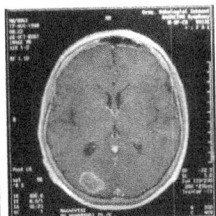

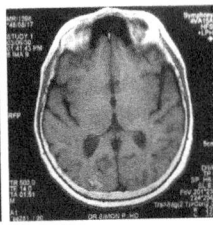

| At the time of the diagnosis, July 2001 | Three months later | Twenty-three months later | Forty-three months later |

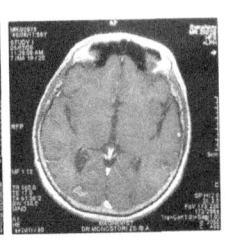

FIGURE 15
Cranial MRI of a fifty-four-year-old female patient at the time of the discovery of a brain metastasis originating from a lung tumor and during the period of DDW consumption.

In another case, a fifty-eight-year-old female patient was diagnosed in late 2007 with a brain metastasis originating from the lungs. The patient started consuming DDW shortly after the diagnosis, and three months later a brain MRI showed significant regression. The patient consumed DDW continuously for the first three years, followed by seven repeated courses of two to four and five months each with several months' interruptions until 2017. The patient died ten years after the diagnosis and two years after the discontinuation of DDW consumption.

The drug Gefitinib, which inhibits growth hormone binding to the EGFR receptor, is known to be effective in 10% of lung tumor patients who have an EGFR mutant gene amplified in their tumor. The cases described above occurred at a time when Gefitinib was not yet used in clinical practice.

Later, in 2012, data from 129 lung cancer cases were statistically processed [54]. The mean age of the 51 female patients was 58.1 years (median: 58.3 years) and the mean age of the 78 male patients was 58.7 years (median: 58 years). 70% of patients (n = 90) were histologically classified as NSCLC, 19% (n = 24) as SCLC and 12% (n = 15) as having a mixed tumor type. Brain metastases were already present in 21% (n = 27) of the evaluated patients. In this publication, we have discussed in detail the survival data of the different histological groups, and the following table shows the survival data at one, two, three, and five years respectively, by sexes (Table 8).

Patients	survival rate			
	1 year	2 years	3 years	5 years
Male	77%	51%	38%	19%
Female	94%	75%	60%	52%
Both sexes	84%	60%	47%	33%

TABLE 8

Sex-specific survival rates at one, two, three, and five years for 129 lung cancer patients evaluated

One surprising finding of the study was the significant gender difference in MST, which was 25.9 months for men and 74.1 months for women ($p < 0.05$).

Basic research results from a mouse model may explain this difference [54], but this needs to be confirmed by further studies. In these experiments, we monitored changes in the expression of the Bcl2, Kras and c-Myc genes in the lung tissue of mice treated with 7,12-dimethylbenz(a)anthracene (DMBA), a chemical carcinogen, twenty-four hours after the induction. The DMBA treatment resulted in a several-fold increase in the expression of the genes tested in female mice in the control group that consumed normal water, with no significant changes observed in male mice. The expression of the three oncogenes was inhibited in female mice given water containing 25 ppm deuterium, confirming previous research findings that a decrease in deuterium concentration has a significant effect on the activity of a large number of genes [53]. In the male mice that were administered DDW 25 ppm after their exposure to DMBA, it did not cause any meaningful oncogene induction, no suppressive effect of DDW could be detected. Whether the significant difference in MST between the sexes observed in the human study is related to this molecular process should be clarified by detailed molecular analysis of tissue samples from human tumors.

After publishing our article in *Nutrition and Cancer* [54], we continued to follow patients with lung cancer and record changes in their condition. At the 3rd International Conference on Deuterium Depletion (Budapest, 2015), we could evaluate data from 304 lung cancer patients. The MST of the patient population studied was forty-eight months, almost six times the expected MST of patients receiving conventional treatment only. The MST values calculated from the data of 129 patients in 2010 and 304 patients five years later are shown separated by sex in Table 9.

	MST of patients enrolled between 1992–2010 (n = 129)	MST of patients enrolled between 1992–2015 (n = 304)	Historical control
Male	25.9 months (n = 78)	40 months (n = 157)	7.5 months
Female	74.1 months (n = 51)	87 months (n = 147)	10.3 months

TABLE 9
Trends in median survival time of lung cancer patients ingesting DDW from 1992–2010 and 1992–2015, by sex

Data from 129 patients followed up between 1992 and 2010 already showed that DDW multiplied the expected survival time of patients, and five years later these values had improved significantly.

Among the lung cancer cases, we highlighted the cases of nine patients who had the tumor removed but continued to consume DDW. The nine patients had a cumulative follow-up of 53.7 years. Two cases of relapse were reported, but no deaths occurred within the examined period. In one of them, the tumor recurred thirteen months after the first surgery and was removed again. The patient then remained tumor- and symptom-free for the following fifteen months. The other patient, after surgery in 2012, repeated DDW consumption for three years, with courses of three months each, with three to six months in between. The relapse of the disease occurred when the patient did not consume DDW for two years from May 2015 until the new tumor appeared. These results are in line with those observed in surgically tumor-free breast cancer patients, confirming that in patients in remission, the tumor-free period can be prolonged and disease recurrence can be prevented and avoided by using DDW.

Colorectal cancer

The third most common tumor type after lung and breast cancer is colorectal cancer. Approximately 12% of the patient population with the potential for complementary use of deuterium depletion can be classified in this group. Several statistical evaluations have been carried out in recent years and the most important results are highlighted.

The study was based on 270 patients diagnosed with colon and rectal cancer who consumed DDW for more than one day. One of the aims of the analysis was to investigate the association of MST with the length of DDW consumption. During data processing, we progressively narrowed the range of patients evaluated. In the first step, we excluded patients consuming DDW for less than thirty days, and then further narrowed the patient population by thirty days to a group of patients consuming deuterium depletion beyond sixty, ninety, one hundred twenty, and one hundred fifty days. The survival data are summarized in Table 10.

Number of cases	Length of DDW consumption (days)	MST (months)
270	1+	84
262	30+	84
249	60+	84
234	90+	84
200	120+	95
177	150+	95

TABLE 10

Variation of MST values as a function of the length of DDW consumption

In Table 10, the MST values of 7.0 years (84 months) and 7.9 years (95 months) for the two groups evaluated are more than twice the MST values of the patient population receiving conventional treatment alone. The calculated eleven months increase in survival for DDW consumption beyond 120 days indicates that a minimum of three to four months of continuous DDW consumption is required to achieve the desired effect.

Further analysis was performed on a homogeneous group of 210 patients who had consumed DDW for more than 90 days and for whom detailed data on disease progression were available. To determine the MST values by stage, patients in stages 0-III according to the classical classification were grouped into one group and patients with stage IV into a separate group. Since the

MST value is significantly influenced by the time between the diagnosis of the disease and the start of DDW consumption, we also examined the evolution of the MST in two further breakdowns. We analyzed separately those who started using deuterium depletion within one and three years of diagnosis, and we also assessed MST separately by sex. The results are presented in Table 11.

	Females			
	DDW within 1 year of diagnosis		DDW within 3 years of diagnosis	
Stage classification	MST (months)	Average (month)	MST (months)	Average (month)
0–III.	–	225.4	–	208.2
IV.	37.8	147.0	42.4	108.8

	Males			
	DDW within 1 year of diagnosis		DDW within 3 years of diagnosis	
Stage classification	MST (months)	Average (month)	MST (months)	Average (month)
0–III.	45.2	81.5	45.6	82.3
IV.	22.1	29.0	30.8	36.1

Table 11
MST values in patients with colorectal cancer, by stage, sex, and time elapsed from the diagnosis to the start of DDW use

In the case of the group of female patients of Stage 0 through Stage III, no MST could be determined due to a low mortality rate in both the group that applied deuterium depletion within one year after the diagnosis nor in the group that applied deuterium depletion within three years after the diagnosis. The MST values in Stage IV patients were double the values of the historical control. In male patients, the MST was higher in all groups compared to patients with the same stage who did not use deuterium depletion, and the increase was more significant in female patients. Given that a similarly significant sex difference was observed for lung cancer, further analysis is needed to clarify what causes this disparity. One reason could be that women are more conscious of their health, recognize the problem earlier and visit a doctor without delay. Efficacy may be also influenced by lower body weight. It is also probable that women follow the instructions more closely and are more disciplined when applying the treatment. Biological differences between the sexes may also be reflected in survival time. It is crucial to further examine and clarify these questions to improve the efficacy of deuterium depletion in men.

Results supporting the anti-cancer effects of deuterium depletion in the entire population

At the 3rd International Congress on Deuterium Depletion (Budapest, 2015), we presented the results of a statistical analysis of data from 1,827 cancer patients who had undergone deuterium depletion treatment. Below are the main conclusions of this analysis.

Patients consented to submit their data between October 1992 and October 2014. All patients who consumed DDW for longer than a day were involved in the study. We had all the necessary data available: age, body weight, sex, date of the diagnosis, the location of the tumor, start and end date of DDW consumption, the status of the patients at the end of the study (deceased/alive), and the changes registered during the study.

56.6% of patients evaluated were female (n = 1034) and 43.4% were male (n = 793). Mean age of patients: 54 ± 16 years (median age: 57 years), mean body weight: 69 ± 17 kg (median body weight: 70 kg). When analyzing the incidence of tumors, it was found that their distribution was nearly identical to the incidence in the entire cancer patient population. Detailed are the major patient groups: tumors of the gastrointestinal tract (18.5%), lung tumors (16.6%), breast cancer (18.7%).

The cumulative follow-up period for patients (from the diagnosis to the end of the follow-up) was 6,881 years. From establishing the diagnosis to the start of DDW consumption, the cumulative period was calculated as 3,104 years. This means that on average, patients started consuming DDW twenty months after diagnosis. The 1,827 patients consumed DDW for a total of 2,265 years. After the completion of deuterium depletion, the cumulative follow-up time was 1,512 years, with a cumulative follow-up time from the start of DDW consumption to the end of follow-up of 3,777 years.

The MST for the entire patient population was calculated as 10.1 years, being the multiple of the MST for patients receiving conventional treatments only. (Also, consider that the number of deaths is generally – albeit with variations across countries – 40–60% of new cases.)

When evaluating the data from homogeneous patient groups, we checked the correlations for the entire patient population by highlighting three main parameters. We examined the evolution of MST as a function of when patients started consuming DDW compared to the time of diagnosis, and also, how the duration of DDW consumption affected the DdU and MST.

Table 12 shows that MST was nearly double (8.1 years) for patients who started to use deuterium depletion within a year from diagnosis, compared to those patients who started it later than two years from diagnosis (MST: 4.5 years). The MST for those who started DDW one year after diagnosis, but no later than two years, fell between the two values, at 6.4 years. Several factors may be behind the significant differences observed. The nearly linear relationship between the diagnosis and the start of DDW consumption and MST clearly shows the decisive role of when DDW as a complementary treatment is fitted to other therapies.

	Time elapsed between the date of diagnosis and the start of DDW (years)		
	Less than a year	One to two years	More than two years
Number of cases evaluated	1,173	209	131
MST (years)	8.1	6.4	4.5

TABLE 12
Evolution of patients' MST as a function of the time between diagnosis and the start of DDW consumption

Table 13 shows the correlation between MST and the duration of DDW consumption, demonstrating a strong correlation between the two factors. Of course, it wasn't just the duration of DDW consumption that determined the

MST of the three groups. The tumor type and stage classification of a given group also determined MST. The 13.3-year disparity (17.9 years − 4.6 years = 13.3 years) between those patients who consumed DDW for three to six months and ten to twelve months, respectively, can only be explained with the significant anticancer effect of deuterium depletion.

	Duration of DDW consumption (months)		
	3–6	7–9	10–12
Number of cases evaluated	512	288	112
MST (years)	4.6	6.1	17.9

TABLE 13

Relationship between MST and the duration of DDW consumption

When investigating breast cancer cases, we also discussed the relationship between efficacy and DdU. There was a demonstrable (inversely proportional) relationship between a given DDW intake and body weight. It was thus investigated whether the above relationship is demonstrable for the entire population of DDW patients. The data (Table 14) are in accordance with previous results, namely, that patients with body weights between 30–60 kg had the longest calculated MST. In terms of DdU, this means that with daily consumption of 1.5 liters of 105 ppm DDW and a body weight of 31–60 kg and 61–80 kg, respectively (a calculated average of 45 and 70 kg), the calculated DdU for the two groups was 1.5 and 0.96, respectively.

	Body weight of patients (kgs)			
	31–60	61–80	81–100	101–
Number of cases evaluated	431	849	327	43
MST (years)	10.9	9.5	9.4	4.6

TABLE 14

Relationship between MST and body weight

The use of deuterium depletion may prevent and protect against the recurrence of the disease

Two other well-defined groups were formed and the data were analyzed. Out of the 1,827 patients, 1,656 had at least one tumor at the start of deuterium depletion, while 171 were tumor-free due to conventional treatments. Key figures from the two groups are presented in Table 15.

	Patients with tumor (n = 1656)	Patients in remission (n = 171)
Cumulative follow-up time from the diagnosis to the end of the follow-up period (years)	6,079	801
MST (years)	9.1 years	Not calculable
The average time elapsed from the diagnosis to the start of DDW consumption (days)	646	374
The average duration of DDW consumption (days)	440	571
Number of deaths	476 (28.7%)	11 (6.4%)
The ratio of cumulative follow-up time/deaths	12.7	72.8

TABLE 15
Key figures of patients with tumors and of tumor-free patients in remission, at the start of DDW consumption

Table 15 shows that patients in remission started deuterium depletion on average 272 days after the diagnosis. The nine-month difference (272 days) suggests that these patients started consuming DDW during or after the conventional treatments, before the recurrence of the tumor.

The MST calculated for the total DDW population (n = 1,827) was 10.1 years for 1,656 patients and those with tumors present, the MST was 9.1 years. During a cumulative follow-up of more than 6,000 years, 28.7% of patients died. It was not possible to establish a control group (no DDW consumption) of cancer patients that precisely corresponded to the DDW population and to compare the data from the two groups. Considering the available statistical data, it is estimated that the consumption of DDW multiplied MST by a factor of four and reduced mortality by over 50%.

The fact that "only" eleven patients out of 171 in remission and consuming DDW died during the 801-year cumulative follow-up period, shows an extremely low mortality rate, compared to the population of patients in the same stage and not consuming DDW. From this low mortality rate, it is reasonable to conclude that in patients in remission, deuterium depletion can prevent the recurrence of the disease by approximately 90%. We should ask the question of whether there is a commonality, and if so, what it is in the eleven patients representing nine different cancers? A detailed analysis of the available data showed that in the cases of two patients with lung and bladder cancers, the short-term consumption of DDW (99 and 91 days, respectively) and in one patient (ALL) an inappropriate deuterium concentration (85 ppm) explained why the desired effect had not been achieved. In four patients (esophageal, colon, uterine cancer, and non-Hodgkin's lymphoma), the duration of DDW consumption was long (1,927, 783, 313, and 189 days, respectively). The patients died years (5.6, 1.2, 2.9, and 3.4 years, respectively) after the discontinuation of DDW consumption when deuterium depletion had no more impact on their system. One patient diagnosed with an astrocytoma grade III brain tumor did not relapse for four years and continued to consume DDW for four years after the relapse. One of three patients with breast tumors was still in remission but started using deuterium depletion 3.5 years after the diagnosis. However, shortly afterward, she developed bone and lung metastases. The other two breast tumor patients started consuming DDW while in remission, immediately after successful treatment. In one case, the disease progressed in the interval between two DDW courses. In the other case, the patient was symptom-free and stopped the consumption of DDW after eight months and resumed deuterium depletion only 2.5 years later, after the appearance of lung metastases.

In seven of the eleven patients who died, the application of DDW was inadequate, which suggests that, even for patients in remission, it is of paramount importance to start deuterium depletion in a timely manner, to schedule the repetition of the regimens appropriately and to determine the optimal duration of the breaks. The basic rules for the use of deuterium depletion are discussed in detail later.

The above conclusions are supported by the fact that the rates have remained almost constant. Of the 2,222 patients evaluated in 2019, 204 started DDW without tumors and 13 patients died during the cumulative follow-up period of 1,024 years (including the eleven cases mentioned above). Of the 204 patients, 156 did not experience a relapse during the 781 years of the cumulative follow-up period. This group of patients was consistently characterized by a pattern of repeated DDW courses of three to four months over years.

Dynamics of disease relapse in light of the recent research

Patients who achieved a cancer-free condition undergo regular check-ups for years to detect any recurrences as early as possible, and to achieve a tumor-free status again with timely treatment. Routine testing methods are in most cases diagnostic imaging (X-ray, CT, MRI, PET/CT) or molecular diagnostic tests such as tumor marker tests. The sensitivity of imaging techniques varies, but even the most sensitive method can only indicate relapse if the size of the recurrence/metastasis reaches the detection threshold of the technology. Tumor marker studies, due to their higher sensitivity, are more reliable in predicting the progression of the disease. Tumor marker tests, however, are not an option for many cancer types, or tumor markers do not provide diagnostic values, and tumors do not produce a higher-than-normal amount of tumor markers. Therefore, this method has its limitations.

A recently developed new diagnostic technique [69, 70] is so sensitive that it predicts the possibility of recurrence and its time course even before the appearance of the tumor by monitoring molecular processes. The technique consists of identifying patient-specific genetic abnormalities in the DNA molecule from the removed tumor. Based on this information, it is possible to track circulating tumor DNA in the bloodstream and its quantitative changes. Two important conclusions can be drawn from the data collected with the new technology:

(1) Patients with no increased number of circulating tumor DNA copies had a 90% likelihood of remaining in remission, without a relapse of the disease.

(2) Widely used diagnostic tools detected the recurrence of the disease during checkups within several months in patients with an increased number of circulating tumor DNA copies.

The increase in the number of tumor DNA copies was detected 8.7 months earlier for colon cancer, 9.5 months earlier for breast cancer, and 4 months earlier for lung cancer. The progression of the disease was also detected with imaging technologies. DNA sequences showing genetic abnormalities specific to cancer cells circulating in the blood (cfDNA, circulating free DNA) are derived from tumor cells present in the body and circulating in the blood. It is common for patients who have undergone surgery, follow-up treatment and are tumor-free, to experience a recurrence depending on the tumor type, but usually months or years after the treatment. The data obtained with cfDNA suggest that distant metastases are a result of tumor cells in the system attempting to survive. In macroscopically tumor-free patients in remission, tumor cells that survived conventional treatments simultaneously engage in cell division and die. Months and years needed for some cells to attach and start growing. This growth is detectable only after a few months or years with conventional imaging. The use of deuterium depletion, with one or more courses of treatment per year, may significantly inhibit the division of tumor cells in patients in remission, thus preventing the recurrence of the disease. The fact that no recurrence was reported in these patients should be associated with low cfDNA levels.

CHAPTER FOUR

Low Deuterium as a Key Element of a Healthy Lifestyle

THE USE OF DEUTERIUM DEPLETION IN HEALTHY POPULATIONS

Apart from a few rare, hereditary disorders, people are generally healthy at birth and the majority of people live most of their lives in good health. Aging is a major "enemy" to our health, as various diseases, mainly chronic, develop with an increasing frequency as we get older. Statistical data suggests that the incidence of cancer is low at a young age, but rises at an exponential rate with increasing age (see: Fig. 2). One of the greatest challenges of our era is how to delay the onset of the disease in an aging population and how to increase the number of years lived in good health. In the next section, we take a look at the basic knowledge we need to achieve this goal, and also, how we can use it and what opportunities deuterium depletion offers in the field of prevention.

The human body is made up of 37 trillion (37,000,000,000,000) cells, with each cell having a genetic code consisting of 3.2 billion letters, in a length of 1.8 meters (6 feet). The genetic code is folded up in the cell's nucleus, the latter being one-hundredth of a millimeter in size. In every cell, at every moment, two thousand biochemical processes are taking place. Thousands of genes are working in a synchronized manner, providing the harmony and balance of which is health.

In cells, there are many molecular processes taking place simultaneously to ensure healthy cell function. However, it is a natural part of life that cells age, change, and sometimes degenerate and turn against the body, becoming malignant. During evolution, certain mechanisms have developed on the level of the cell and that of the system to recognize and eliminate cells that threaten the survival of the organism. One of these mechanisms "counts" the division cycles a cell has undergone. Once having completed around fifty cell divisions, cells trigger a self-destruct mechanism called programmed cell death, or apoptosis. This mechanism is designed to prevent the survival of the cell that would inevitably accumulate and pass on more and more genetic defects with each future division. In addition to the molecular protective mechanisms of cells, the immune system also plays an important role in preventing cancer. Although a high incidence of cancer shows that the immune system is still not effective 100% of the time, allowing tumor cells or tumor cell groups to form and survive in any part of the body. These can then keep growing unnoticed in the body for years, so by the time a tumor is diagnosed, four to five years may have passed since the appearance of the first cell. Screening is designed to detect and destroy these tumor cells at the detection threshold at an early stage, but the controversies and dilemmas surrounding screening reflect the challenges.

One approach attributes the development of cancer to mutations or other errors in the genetic program. It has been studied and confirmed decades ago, and recently re-confirmed by PCR (Polymerase Chain Reaction) technology, that the presence of deuterium increases the number of mutations and genetic defects [71]. Observations confirmed by experiments are supported by quantum physics considerations. It is well known that, according to Watson and Crick's model of the DNA, the two DNA strands are reinforced by double and triple hydrogen bonds between the A–T and G–C base pairs (A: adenine, T: thymine, G: guanine, C: cytosine). The movement of the hydrogen atoms involved in the formation of the hydrogen bonds is harmonic concerning each other and is determined by the vibrational frequencies of the bond. Conversely, if one of the hydrogen bonds is replaced by a deuterium bond, where the vibrational frequency of deuterium is about 30% different from the vibrational frequency of the hydrogen in the hydrogen bond, the bond strength of that base pair may be significantly weakened because the harmonic motion of the hydrogen bonds is eliminated. The presence of deuterium may significantly affect the accuracy of the replication process during DNA synthesis due to the difference in chemical behavior resulting from its double atomic mass, as confirmed by PCR experiments [71].

Based on the above, it is reasonable to assume that a lower deuterium concentration during deuterium depletion also reduces the number of mutations, and this may be one of the mechanisms by which DDW may contribute to the prevention of cancer.

It is hard to imagine that a relatively small reduction in the body's deuterium concentration can have a substantial biological effect. This is illustrated by showing the absolute amount of deuterium involved in the formation of hydrogen bonds during a person's lifetime.

In a person's lifetime, about 10^{16} cell divisions take place in the body. Before cell division, a cell needs to duplicate 1.8 meters (6 feet) of genetic information. Therefore, the total length of the genetic program that our cells synthesize is 18 trillion kilometers, with one deuterium bond every 0.0014 millimeters within a DNA molecule, totaling about 1.2×10^{22}. It is clear from these values that reducing the body's deuterium concentration by even 10% will reduce the number of deuterium bonds by an absolute value of 10^{21} during DNA replication in cell divisions over a person's lifetime.

It is important to stress that mitochondria, the powerhouses of the cells, have their own genetic makeup. Mutations in the mitochondria also affect cell function. It is generally accepted that properly functioning mitochondria are a prerequisite for healthy cell function. However, the discovery by Nobel Prize-winning researcher Otto Warburg in the 1920s shed light on the fact that cancer can be caused by alterations in the mitochondrial breakdown metabolism, which can be traced back to mutation(s) in the mitochondrial genetic code.

One general conclusion is that the consumption of DDW and a life spent maintaining a lower deuterium concentration can both contribute to the preservation of health and prevention of cancer by reducing the number of mutations. At this point, the presence of deuterium and its effects explain both the genetic and metabolic approaches to cancer.

Despite the above, we still must accept that even if it's possible to reduce the number of genetic defects and thus the number of cancer cells, it is not one hundred percent effective in preventing cancer. Concerning prevention, an important challenge is whether the treatment is capable of destroying the tumor cells before they reach the diagnostic threshold and/or form distant metastases. This underlines the importance of prevention, as in fact, "cancer patients" are cancer patients well before the diagnosis. All diagnosed cases were already considered to be cancer years before, but the disease only reached the detection threshold later. This also means that potentially everyone is constantly producing tumor cells or tumor cell groups, and there is no way to tell how many months or years before the disease is confirmed. (*The mechanism of cancer development is discussed in Chapter One.*)

The cancer-preventive effect of deuterium depletion is supported by several experimental results. Thanks to the rapid development of diagnostics, it is now possible to follow precisely the physiological and biochemical processes taking place in the human body. One such physiological parameter that can be monitored is the tumor marker. In the case of certain tumors, molecules – proteins mostly – appear in the bloodstream, urine, or tissues that are either produced by the tumor itself or by the body's response to the tumor. Elevated tumor marker levels, once diagnosed, can be used to monitor the progress of the disease and the success of treatment, as their levels, in most cases, correlate well with the size of the tumor present in the body. This is demonstrated by the fact that high tumor marker values measured before surgery are drastically reduced after surgery.

In some cases, even if the surgery is successful, the tumor marker may rise again sooner or later, even if imaging tests (X-ray, CT, MRI, PET/CT) cannot detect a new tumor in the body. (As described previously, tumor marker levels can also be increased due to inflammation, so a high tumor marker level alone has no diagnostic value, and can only be used in planning a deuterium depletion treatment if there is a clear relationship between the presence of a tumor and an increased tumor marker level.)

The efficacy of deuterium depletion is well-monitored in patients in whom imaging technologies have not yet detected a tumor, but whose tumor markers have already increased. Tumor marker values continue decreasing after the consumption of DDW. In another group of cases, the effect of deuterium depletion did not lead to an increase in tumor marker values, even though this happened repeatedly in previous years.

Cervical cancer is a type of cancer that is one of the most sensitive to deuterium depletion and one of the most curable. One reason for this is its anatomical location, which allows the affected area to be examined directly with a colposcope (a long-focus lens microscope with a magnification of 25–50x) without invasive intervention, and cytologists can categorize the histological sample with a microscope after taking a swab from the cervix. The staining method for smears (the "Pap smear") was developed by Georgios Papanicolaou, and his work is the basis for the introduction of the P0–P5 scale. P1 and P2 indicate no abnormal processes in the cervix, P3 indicates the presence of an abnormal lesion, and P4 and P5 indicate a cancerous lesion of the cervix.

In the case of the cervix, it is possible to directly follow the process over several years from a healthy to a cancerous epithelium. The anti-cancer effect of deuterium depletion is demonstrated by the fact that the pre-cancer P3–P4 stages were reversible in many cases. As an impact of DDW, repeated examinations showed P2 cytology findings, without the need for intervention.

Directly investigating the preventive effect of deuterium depletion would require following thousands of people taking DDW for at least a decade, which was not possible. However, we were able to do a follow-up of successfully operated tumor-free patients, who were several times more likely to have a tumor recurrence than the healthy population. In this population, a low rate of regression and death rate point to the high efficacy of DDW. Long-term animal studies and more than two decades of experience with Vetera-DDW-25 antitumor veterinary medicine have shown that a course of treatment (1.5-2 months per year) significantly reduces the risk of disease recurrence, in those small animals made already tumor free.

Of the 48 patients in remission who were previously diagnosed with breast cancer and operated on, only one patient died during more than the 221 years of cumulative follow-up of patients taking DDW [65]. Extending this analysis to cases starting DDW consumption in remission followed between 1992 and 2018, similar results were obtained.

Of the combined nine lung cancer cases, thirty-two gastrointestinal, fifteen gynecological, and eight prostate tumor patients in remission, only five patients (four gastrointestinal and one gynecological) died during the 338-year cumulative follow-up.

Based on the above, we recommend deuterium depletion for healthy people as described in the *Dosing advice* section. Consuming DDW is recommended as a preventive measure for individuals aged forty to fifty in the form of courses (with deuterium levels of 125 ppm or 105 ppm), every two or three years. For people over fifty or in a high-risk group (smokers, people with a family history of accumulation, etc.), a course is recommended every one to two years, according to *Protocols H/1* and *H/2*.

To summarize the above, healthy people can increase the number of years they live without disease if they can keep their body's deuterium concentration 10-20 ppm lower than the average of 145-150 ppm, and occasionally take a course of treatment to reduce this deuterium concentration even further over a few months.

The limitations and contradictions of screening tests

In addition to the general dietary and health-promoting lifestyle advice, the use of screening offers an additional opportunity to reduce cancer mortality. The case for cancer screening is supported by experience to date, which shows that the earlier the disease is detected, the greater the chance of a complete and permanent cure. A review of the results of cancer screenings to date also shows that there are some limitations to its effective implementation in practice:

1. There are more than two hundred different types of cancer, only some of which can be screened by mass screening.

2. The sensitivity of screening methods varies. If a tumor in the person being screened was below the detectable level, it may not have been detected. However, if the test had been carried out a few months later, the patient could have been successfully screened.

3. It also follows that a negative result at a given moment in time only temporarily confirms the absence of tumors.

4. By screening the population for only one type of tumor, while those particular findings may be negative, another type of tumor may be present and undetected in the person being tested.

5. The widespread use of state-of-the-art imaging (CT, MRI, PET/CT) to perform whole-body screening is pointless, impractical, and economically unsustainable.

6. Any screening is only effective if it succeeds in screening at least 50-60% of the target population, which screening campaigns usually fail to do.

7. The basic requirement for screening is a high degree of redundancy. Since we do not know in advance who is sick, we screen „everyone" in the hope that there are some people who have a tumor detected in the early stages of the disease. The magnitude of the challenge is illustrated by the fact that, for example, for mammography examinations, one out of every 1,700-1,800 tests is positive.

8. It is often the case that a false-positive result is obtained, which places a considerable psychological burden on the person who has been screened until it is clear that a misdiagnosis has been made.

9. While it is known that cancer detected at an early stage can be treated with better results and a higher chance of being cured, it is impossible to say for a given patient what his or her life expectancy would have been if treatment had been delayed until years later.

10. A further dilemma is that proving the presence of tumor cells does not necessarily mean the patient will develop a tumor. A significant percentage (60-70%) of elderly men who have died from non-prostate tumors have tumor cells in their prostate at autopsy. This further complicates the question of whether treatment is needed and, if so, when should it start?

11. There can be serious health, economic and social consequences if a large number of patients are identified with cheap, simple screening for cancer cells, followed by treatment with significant costs. The case of PSA (prostate-specific antigen) tumor marker testing illustrates this problem. PSA is elevated in a significant percentage of the elderly male population, and there is a fifty-fifty chance this can be attributed to either inflammation or prostate tumors. The difficulty is that it is not possible to predict who will develop a tumor, and treatment of all men with high PSA is economically unfeasible.

Partly due to the difficulties of screening, one approach to cancer prevention was to give healthy people with a high risk of breast cancer the anti-cancer drugs used in oncology. Experience has shown that, although there was a demonstrable reduction in breast cancer in the population followed, the preventive treatment was accompanied by serious side effects and life-threatening events (e.g. pulmonary embolism) due to the toxic effects of the drug. Given the results, it was recognized that the drugs used in cancer treatment cannot be used to prevent cancer in healthy people because of their toxic effects.

The use of deuterium depletion in medicine could also bring a breakthrough in the field of prevention, as it is currently known to have an anti-tumor effect without causing toxic side effects.

CHAPTER FIVE

Use of Deuterium Depletion in Benign Tumors

There were also a limited number of patients with benign tumors among those using DDW. The results and experience were mixed in this disease group. In some cases, there were noticeable positive effects of deuterium depletion, while in other cases no significant changes were observed. In most cases, the benign tumors did not show the same degree of sensitivity as observed in the malignant tumors. Therefore, there seems to be a clear difference in sensitivity. While healthy cells did not show sensitivity to deuterium withdrawal, benign lesions showed a lower degree of sensitivity, which was more pronounced in malignant tumors.

In the case of benign lesions, despite their low sensitivity, it is recommended that patients should consume DDW (in the range 105-85 ppm) for three to four months, during which time it can be determined whether further improvement can be expected.

Another approach to the problem is for an individual with a benign disease to consume DDW according to the H/2 Protocol, with the secondary goal of preventing the benign tumor from developing into a malignant tumor.

A good example of the latter is the case of a twelve-year-old girl who was diagnosed in 2000 with an astrocytoma grade I tumor in her spine, which filled a long stretch of the spinal canal. No conventional therapeutic options were available to treat the disease. The patient had been consuming DDW on a more or less regular basis for many years, and an MRI scan in 2010 showed that the tumor had decreased in length by 2 cm compared to the scan in 2002. In 2015, the patient graduated from college and was later symptom-free in 2018 as well.

In patients diagnosed with meningioma, deuterium depletion may be of importance after successful surgery to prevent or significantly delay the recurrence of the tumor.

For fibroids, minimal sensitivity was observed when DDW was used, based on data from a limited number of patients.

For men with frequent benign prostate enlargement, a concentration of 105 ppm is recommended, which may relieve urinary symptoms, but more importantly, with repeated DDW courses the development of malignant prostate tumors can be prevented.

The appearance of cysts has been linked to the malfunction of the fumarate hydratase enzyme, which plays an important role in the mitochondrial respiratory chain (Szent-Györgyi-Krebs cycle). Therefore, the fact that studies with DDW have shown that lower D concentrations improve mitochondrial function may be of particular importance in the treatment of cystic diseases. Further research is needed to clarify this, but the high sensitivity of the cysts and the rapid response of patients is attributed to either the destruction of cells containing malfunctioning mitochondria or the improvement in mitochondrial function by DDW. Benign cysts are sensitive, generally responding well to deuterium depletion, with rapid and significant size reduction or complete resolution of cysts in a relatively short time frame.

CHAPTER SIX

Deuterium Depletion as a Supplementary Treatment for Malignant Tumors Used Alongside Oncological Treatments

A Paradigm Shift in Curing Cancer

Including DDW among resources available for cancer treatment does not just mean merely adding a new pharmaceutical substance to the toolbox, but also promoting this new medical strategy. The primary goal of most tumor treatment strategies put into use in the past several decades, one can see that the primary goal has been to destroy tumor cells using cytotoxic compounds and other medical procedures. Cancer research and drug development have focused mainly on finding molecules that inhibit cell division and destroy rapidly dividing cells. However, rapid cell division takes place not only in tumors but also in several other parts of the body. This means that these treatments have severe and visible adverse effects. In many clinical studies, a moderate increase in efficacy against tumors coupled with slightly reduced adverse effects is considered a good result, sufficient for the new substance to be recognized as a "better" drug than those already in use. Oncology involves balancing the goal of producing drugs with ever-higher efficacy against tumors, while not allowing adverse effects to exceed tolerable levels. Cancer death statistics unequivocally indicate the failure of this approach. Acknowledging and realizing this fact, new advancements in molecular biology and the unsatisfactory results of the targeted therapeutic approach have given rise to the goal of "taming" cancer, converting it from an acute disease to a chronic one. This realization is based on the fact that it is not the approximately one- to two-centimeter tumors that causes the deaths of cancer patients. The real cause is that said tumors continue to grow and interweave with the adjacent tissues, and create metastases, all of which collectively cause significant physiological changes and the deterioration and loss of crucial bodily functions, ultimately resulting in the patient's death. The approach of "rendering the disease chronic" renounces the objective of destroying tumors. Instead, its main goal is to keep the mass of tumors under control on a preferably constant level, while achieving all this with minimal toxicity and maintaining the patients' general quality of life to the greatest extent possible. This approach can contribute markedly to the long-term survival of cancer patients by allowing the use of new drugs with mild adverse effects and the use of existing drugs at a reduced dosage.

The supplementary use of deuterium depletion may significantly improve the effectiveness of modern oncotherapy, for various reasons.

The combined use of deuterium depletion and conventional treatments leaves a significant percentage of patients completely cancer-free. Using DDW as a follow-up treatment after a conventional regimen helps to reduce the risk of remission.

In patients who cannot be completely cancer-free, the supplementary use of DDW alongside conventional treatments may reduce tumor size and increase patient life expectancy.

Patients may develop resistance to medications, including DDW. Combining, fitting, and synchronizing different procedures can prevent the development of drug resistance.

Only a limited number of conventional therapeutic methods can be used, and each one can only be used for only a limited period. However, DDW can be administered with minor or even no interruption for very long periods. In this way, it plays an important role in preventing the disease from recurring and keeping patients' cancer-free. Deuterium depletion makes it possible to not intervene only if the disease should recur, but also provide active treatment for patients already in remission.

One basic goal of the use of deuterium depletion as a supplementary treatment is to find and adequately ensure the balance between the severity and efficacy of the treatment and the stage of the disease, even over many years. It also establishes how available resources must be allocated to combat the disease. One main aspect is that both conventional treatments and deuterium depletion should be used only to the extent necessary. If a given medical procedure, for example, chemotherapy, is efficacious in itself, then we might consider simultaneously using complementary treatments. It is also crucial to take advantage of the additive and synergistic effects of DDW and conventional treatments.

Changing the concentration of deuterium also adds to the many uses of DDW. It is not necessary or recommended to start with the lowest deuterium concentration right away. A moderately reduced deuterium concentration value, closer to the one found in nature, can also achieve the desired effect when combined with conventional treatments.

Several factors must be taken into consideration when planning treatment. The stage of the disease and tumor pathology are decisive factors. It is also crucial to determine which conventional treatments are suitable, their expected efficacy and whether the patient can be brought into remission. Other factors include the degree of improvement expected if conventional treatments do not result in the patient becoming completely cancer-free and the expected period without disease progression. It is important to know how the DDW regimen start date is related to the time of the diagnosis. There is a substantial difference between beginning deuterium depletion at the time of the diagnosis and using it as a supplemental treatment for a progressive disease that has been ongoing for years.

Considering different treatment options and their combinations, the progression of treatment, and the various ways the tumor may react, it is difficult to come up with a single sequence of explanations that cover all the questions and tasks related to the use of DDW alongside conventional treatments. We, therefore, consider many aspects of this problem.

A comparison of conventional and submolecular treatment strategies

To compare and give a summary of the above two treatment strategies, it should be noted that while conventional treatments aim to destroy tumor cells, deuterium depletion aims to "tame" them. The two strategies are governed by entirely different principles and require different mindsets.

A few basic differences:

Conventional treatment	Deuterium depletion
Non-natural active substance	Nature-identical active substance
Targets a specific point of cell division	Affects the regulatory mechanism of cell division and metabolism in multiple points
Has serious, toxic side effects	No harmful side effects
One drug is generally applicable for a specific type of cancer	Applicable for a wide spectrum of cancers
Aims to destroy the tumor	Aims to reprogram the metabolism of malignant cells, restore a healthy mitochondrial function and restrict the functioning of malignancies

Despite the fundamental differences between the two treatment concepts, the goal is to combine these treatments to provide patients a better chance for a complete recovery, prevent relapse, and substantially extend the time spent without disease progression, while maintaining a good quality of life.

The most basic goal is the same in both cases. First, a completely cancer-free condition of the patient must be achieved (meaning that no tumors should be detected in the body by using imaging technologies). Subsequently, the cancer-free condition must be maintained, preventing the recurrence of the disease. (In terms of the treatment outcome, it is indifferent whether conventional treatment methods or a combination of DDW and conventional treatments are used. The outcome is determined by the stage a patient was in when the treatment[s] started.) Patients are categorized into three stages, irrespective of the type of cancer:

- Stage I: A well-circumscribed, operable, non-infiltrating tumor, 1–2 cm in size, no metastasis in the surrounding lymph nodes.
- Stage II: Tumor size exceeds 2 cm, may infiltrate the surrounding tissues, tumor cells are detectable from the surrounding lymph nodes, inoperable, but may be made operable with proper treatment.

- Stage III: Distant metastases are detectable with medical imaging independently from primary tumor size; inoperable tumor, or even if surgery is feasible, it only produces minor results, such as reducing tumor size and the number of tumors, with patients not becoming cancer-free.

Conventional treatments can render patients cancer-free in the majority of stage I and stage II cases. Results can be achieved by using surgery, radiotherapy, chemotherapy, and hormone therapy. In the event of a patient already having distant metastases (stage III), a completely cancer-free condition cannot be achieved. This means that with conventional treatment alone, the primary goal cannot be achieved. The combined use of conventional treatments and deuterium depletion helped to achieve a cancer-free condition in a significant percentage of stage I and stage II patients, and in a higher percentage than is currently the case for stage III patients. The combined use of conventional and submolecular treatment strategies makes it possible that a higher percentage of patients can be rendered cancer-free. As part of aftercare, the combined use of both treatment methods can prevent the recurrence of cancer. Also, it is possible to halt the progression of cancer in patients who could not achieve a cancer-free condition.

Deuterium depletion may influence tissue pathology

Prior treatment of pets shows that the effectiveness of DDW is not limited to just causing the death of tumor cells and reducing the size of the tumor, but it is also capable of influencing the malignancy of the tumor or even reversing the process of tumor formation. The pathology of samples taken from different phases of the treatment from a dog suffering from rectal cancer confirmed the morphological transformation of tumor cells. The histology of the tumor described before the DDW treatment as a typical adenocarcinoma changed in the sample taken after a few weeks of DDW treatment. Cells became more uniform and differentiated and showed the trait of benign adenoma. In a new sample taken a few weeks later, no tumor cells were detected, but only a strong lymphoid infiltration indicated increased immune activity (see Figs. 16–18).

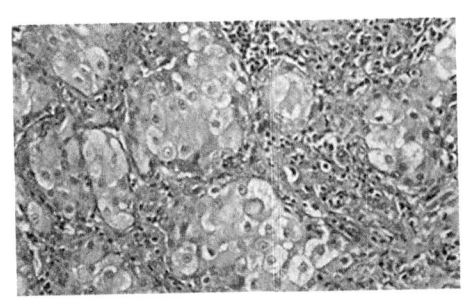

FIGURE 16
Histological image of a fourteen-year-old male spaniel's rectal tumor before the start of the treatment with Vetera-DDW-25; *oncocytic carcinoma*

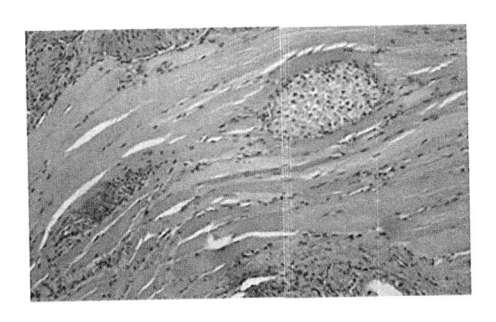

FIGURE 17
Histological image of a fourteen-year-old male spaniel's rectal tumor after four weeks of treatment with Vetera-DDW-25; *apocrine adenoma*, cells became smaller and more uniform, the degree of differentiation of tumor cells increased

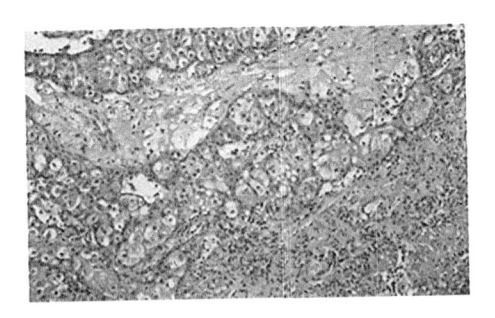

FIGURE 18
Histological image of a fourteen-year-old male spaniel's rectal tumor after eight weeks of treatment with Vetera-DDW-25; strong lymphoid infiltration in the sample from a previously tumor-affected region

Such effects of deuterium depletion are confirmed by reports cited in the chapter, *"Low deuterium as a key element of a healthy lifestyle."* The cited reports show that the cytology results of female patients consuming DDW changed from P3/P4 to P2.

The effects of deuterium depletion on pathology were confirmed by cases in human and veterinary medicine. Here, DDW was applied before surgery and the pathology of the tumor removed during surgery was better than expected, based on preliminary cytology tests. In other cases, a large number of apoptotic tumor cells were found in the removed tumor. There was also a reported case when the changes were so significant that it was impossible to establish an accurate diagnosis. There was one case when the bone sample from a patient with Ewing's sarcoma did not exhibit any structure. Tumor cells underwent lysis in response to deuterium depletion.

The above results raise the question of whether it would be beneficial in the future to let pathologists know of the fact that the examined sample is from a patient who had consumed DDW. This would provide a clue for the correct interpretation of the varying, and at times confusing, histology results. In the coming years, the collection of similar data may provide valuable information for establishing the correct dosage of DDW and for a better understanding of its mechanism of action.

The dilemmas of histological sampling

Heisenberg's uncertainty principle states that it is impossible to determine the location and angular momentum of electrons (subatomic particles) at the same time. The more accurately one parameter is determined, the less accurate the other parameter becomes. This phenomenon is an inherent property of quantum systems and it gives rise to the "observer effect." The observer effect means that observing a situation or phenomenon necessarily changes it.

A basic principle in oncology diagnostics and planning treatment is that tumor cells must be extracted and the sample must be examined with the tools of cytology and histology. The treatment cannot even begin without this examination, as the findings from a pathological examination provide the strongest evidence for a precise diagnosis. Doctors "jab" into the tumor with a thin or a thicker needle a number of times, if needed. Sampling with a needle damages the structure of the tumor. Blood vessels may be ruptured in the process, causing bleeding or inflammation, not to mention the breaking off of tumor cells, and their drifting to distant parts of the body through the bloodstream. While determining the tumor's histology, medical interventions are performed on a patient whose primary tumor has been biopsied.

The "observer effect" analogy from quantum physics fits this routine practice. While one of the critical issues in intra- and postoperative pathology evaluation is determining whether a vessel has already burst, which impairs the patient's chances of recovery, this is deliberately induced during the biopsy to obtain a diagnosis. It breaks down the structure that has been formed over the years as a result of the struggle between the tumor tissue and its healthy surroundings, and which has served to allow the body to localize the tumor.

The dilemmas surrounding the practice of histological sampling are further complicated by the observation that 60–70% of the prostates of the sixty- to seventy-year-old male population who have died of non-prostate tumors had tumor cells present. If these patients had been tested for PSA tumor markers, they would most likely have shown a higher than the normal range, even if the death of these patients was not caused by prostate tumors. The uncertain question is in which of the many potential patients cancer will manifest and why. If a high PSA level is detected, the doctor performs a biopsy, sometimes followed by total removal of the prostate (if the cancer is operable and the patient agrees to surgery), saying that the tumor should be removed while it is still in its capsule. (The tumor's capsule is punctured multiple times in search of tumor cells in the prostate.) PSA levels drop following a surgery and can be maintained at a low level for a long period by using radiotherapy and hormone therapy. However, PSA can still increase later, and at that point the options for intervention are limited.

If patients are diagnosed at a stage when they are no longer operable, they will receive the above therapies, but before this, tissue samples are also taken from the patients.

Doubts related to biopsies are reinforced by the fact that one cannot conclude that a patient is healthy even when tumor cells are not detectable in the sample. There is a possibility that even though there were no tumor cells at the sampling location, tumor cells could be present in any other location of the biopsied organ. In this case, only the examined area is damaged, but no treatment is initiated.

Doctors try to resolve this contradiction by not intervening immediately, despite a high PSA value, but by regularly monitoring the tumor marker levels with a so-called watchful waiting. No cancer can be treated without a histological diagnosis, so the best thing to do is just stand by and wait, hoping that the patient remains in a stable condition for up to several years. If results indicate a further increase of PSA, the expected medical interventions are applied automatically (biopsy, surgery, radiotherapy, and hormone therapy). Conversely, if tumor marker values do not increase (or they do decrease), then the patients have been spared an unnecessary intervention or the complication arising from a biopsy.

The integration of deuterium depletion may significantly transform current practices and help resolve the controversy surrounding biopsies. If clear signs are pointing to the presence of a tumor in the body, then the changes as a result of deuterium depletion are indicative. Decreasing tumor marker levels or the reduction of the tumor mass (detected by imaging technology) may be related to the tumor necrosis induced by the consumption of DDW. This may open a new chapter in the treatment of cancer, because although some information is not available without a biopsy, deuterium depletion may start reprogramming the cellular metabolism of the cancer patient without interfering in a far-reaching and invasive manner and disturbing the structure of the tumor. After a successful surgery following a regression, a detailed histological examination may help detect other, cancer-related properties (receptor status, degree of differentiation, the expression of oncogenes, etc.). All these properties can and must be taken into account when planning a treatment regimen.

We discussed earlier the case of a prostate cancer patient who was biopsied and diagnosed with a tumor in 2009. The patient did not consent to surgery and conventional treatment but started consuming DDW. The patient first observed a nine-month drinking cure, followed by a four to five-month one. He then switched to three-month courses, with three to four months of interruption. Four years later, the previously 1 cm large tumor could not be detected anymore. After three years (seven years after the diagnosis), the doctor deemed it necessary to repeat the biopsy which produced a negative result. The reason for this was a contradiction in that the patient had no symptoms for seven consecutive years, even though he did not receive any conventional treatments.

The negative histological results confirmed that the patient could have avoided an invasive treatment. This is further confirmed by the fact that when the patient stopped consuming DDW for eleven months, PSA levels started to increase again, and nine years after the diagnosis, the MRI showed the presence of the tumor again. So, even if the histological findings were negative, this did not guarantee a cancer-free condition. This again confirms the fact that two conclusions can be drawn from negative cytology results: either the patient is healthy, or the sample was taken from the surrounding tissues where no tumor cells were present.

Another patient was diagnosed in 2007 with a breast tumor with bone metastasis. The patient underwent surgery and received protocol treatment, while also taking the option of deuterium depletion. Consecutive check-ups over years detected the metastasis, but without any change. Between 2007 and 2012, the patient repeated DDW treatments for three to four months with three to six and later eight-month pauses, and then did not continue deuterium depletion for eighteen months after the end of 2012. Subsequently, in the summer of 2014, checkups showed a one-millimeter increase of lump in the vertebra. Even though the patient was diagnosed with breast cancer and received conventional treatment, the lesion seen in the bone was not considered to be a breast cancer metastasis at that time, as the bone metastasis would have progressed over seven years. Subsequently, a bone biopsy was performed at the patient's request, which confirmed that the metastasis was indeed a malignancy of the breast in the vertebra and had stagnated for many years.

These two cases illustrate that the additional use of deuterium depletion may significantly influence the time course of cancer progression. The fact that the condition of the breast cancer patient diagnosed in stage IV did not deteriorate over five years does not mean that the results of the checkup were wrong. It only means that the progression of the disease may be halted by using deuterium depletion.

The alignment of deuterium depletion with imaging tests, their sensitivity and influence on the results of the tests

It is a common phenomenon that the necrosis of tumor cells due to deuterium depletion is accompanied by an inflammatory process. Thus, the tumor size may initially increase during treatment and the patient may experience a warming of the affected area. This has also been confirmed by CT, MRI, and PET/CT scans, the results of which can also be confusing if the evaluating physician is not aware that the patient is consuming low-deuterium water. Studies do not provide representative results for approximately three to four weeks after the beginning of a DDW treatment or when increasing the dosage (ideally, two to three months after the start of a DDW treatment or switching dosage). In this case, the apparent growth in the size of the tumor is likely due to an inflammatory process, thus the confusing results.

If the consumption of DDW is interrupted, it is recommended to schedule the imaging test at the start of the interruption. It may also offer new insights to perform an imaging test after several months of interruption, but before resuming the consumption of DDW. A comparison of the results may reveal whether the disease has progressed during the interruption of deuterium depletion. If the disease has progressed, it means that the pause was either premature or lasted longer than ideal. If there has been no change, it indicates that the timing of the interruption was correct. In that case, the duration of the next interruption should be extended by one, two, or three months.

PET/CT studies offer an opportunity to obtain information simultaneously on the metabolic activity of a given area (PET) and its anatomical structure (CT). During PET scans, radioactive isotope-labeled molecules, usually sugar (^{18}F-fluorodeoxyglucose, FDG), are administered to the patient, which, due to more intense metabolic processes in tumor cells, more strongly marks them. (In some cases, areas of inflammation are also more strongly marked.) *In vitro* experiments with sugar molecules labeled with C^{13} (the heavy isotope of carbon) have shown that cells kept in a growth medium with DDW have increased glucose uptake, taking up 15–20% more sugar. No studies have been planned to confirm whether patients consuming DDW are more intensively marked in tumor areas, but the possibility has been raised in some routine examinations. This suggests that a few days of DDW consumption before a PET/CT scan may increase the sensitivity of the examination. Consuming DDW supposedly causes an increased sugar uptake in the stimulated cells, rendering cell groups that could otherwise not be traced visible. Thus, a sensitized PET/CT scan may reveal a metastasis that potentially cannot be detected without stimulation with DDW.

Deuterium depletion may affect tumor marker levels

Lower costs, simplicity, and more frequent testing are the advantages of tumor marker tests over imaging tests. A further advantage is that changes in tumor marker levels indicate the progression of the disease at a point when they cannot be detected by imaging tests. Concerning tumor marker tests, the procedure is similar to that of imaging tests. Tumor necrosis and the accompanying inflammatory response to deuterium depletion may temporarily increase tumor marker values. Therefore, tumor marker testing is not recommended shortly after the start of DDW consumption or after the increase of DdU, but rather after two to three months.

In some cases, the diagnostic value of tumor markers is limited, so, understandably, they are not given too much weight in the presence of a definitive histological record. However, in terms of the long-term follow-up of patients, there is an indication for tumor marker testing even if it is not crucially important for the diagnosis. This is important mainly to reveal whether there is a relationship between the presence of the tumor and the tumor marker levels. Given that tumor marker levels do not exceed the normal range in all cancer patients, this question can only be answered if the tumor marker levels are determined at the first checkup (before surgically removing the tumor and administering a conventional treatment to the patient). If the test is not done in time, then the appearance of normal tumor marker levels months later is uncertain. If it does happen, it can mean that the disease is not progressing. It can also mean, however, that the cancer is progressing, but without increasing tumor marker levels.

Deuterium depletion may multiply the life expectancy of patients and may lead to complete recovery in a high percentage of cases. This makes it necessary to follow the patients in the long term. Naturally, this requires intermittent CT, MRI, or PET/CT testing. However, monitoring the tumor marker levels (once their diagnostic value is convincing) provides a solid foundation for the monitoring of patients and establishing the dosage.

The correct definition of DdU and the principle of dosage

The modus operandi of DDW and the principle of dosage is quite different from what is customary in conventional therapies. It is widely known that the dosage for chemotherapy medicines is calculated and applied based on body weight and body surface area. The dosage may also vary depending on how patients respond to treatment. The guiding principle, however, is that a constant amount of cytostatic agents should be used following internationally accepted protocols.

On the other hand, the deuterium concentration of DDW will change during the treatment, as the goal is to maintain a constant change in deuterium levels. The change should reach the level required for the treatment to be effective and it should be maintained as long as possible. If it is not possible to reduce deuterium levels anymore, the goal is to keep the deuterium concentration at the lowest possible value.

The greatest reduction in deuterium concentration in the patient's body may occur in the period after the start of DDW consumption. As the body's deuterium concentration decreases during the application of DDW, the same daily dose results in a smaller and smaller decrease in deuterium levels. To ensure a daily reduction in deuterium concentration, after a certain period (approximately one to two months), it is recommended to continue the treatment with a DDW concentration of 10–20 ppm (preferably 20 ppm) lower than what was used previously. At this point, the concentration gradient between the patient's body and the lower-deuterium DDW is re-established, which also leads to a further significant decrease in deuterium concentration. Ideally, it would be possible to adjust the deuterium concentration of DDW daily, but experience has shown it is sufficient to change or reduce it every few weeks or everyone, two or three months. Thus, a concentration gradient between DDW and the deuterium concentration in the body's fluid compartments can be maintained, even if the deuterium concentration in the patient's body substantially declines in the meantime.

After the initial DDW dosage has been established, it is important to follow and analyze the changes. The effects of DDW are detectable within a short period, and in the vast majority of cases, the preparation induces a rapid, noticeable response. This manifests itself in rapid necrosis of the tumor, which generates a complex response from the body. Inflammatory processes can be triggered in the necrotizing areas, necrotized cells need to be processed and this can put a strain on the body as a whole. The reason for this is that processes such as tissue structures are being remodeled. Necrosis may cause pieces of tissue to detach from the tumor, and the affected area (colon, bladder, lungs, etc.) needs to be restructured. One of the most important rules for the application of deuterium depletion is that the healing process must not be hastened. The various recommendations for use have been compiled with this in mind.

CHAPTER SEVEN

The Application of Deuterium Depletion before Diagnosis up until the Start of Oncological Treatments

Indicative Signs before Diagnosis

It is characteristic of only a few cases in oncology that a disease manifests itself overnight with complete certainty. Typically, it takes four to five years for a tumor to reach a size that causes noticeable symptoms. It is at the end of this period that patients realize that something is wrong with their bodies. This is one of the most critical stages in the later course and treatment of the disease, because if managed well, the patient's chances of survival may improve significantly. Unfortunately, experience has shown that mistakes are often made at this stage, which can make life difficult for both patients and doctors.

When unusual symptoms such as weakness, a rapid loss of weight, persistent cough, bloody stools, lumps, difficulty swallowing, frequent urination, double vision, or other symptoms appear, patients should consult a doctor as soon as possible and request a thorough examination.

Spontaneous healing is not characteristic of cancers. It is very rare to have a palpable tumor grow and then suddenly disappear from the body. Only one out of ten thousand patients are lucky enough to have such a tumor. No one can expect a spontaneous disappearance of the symptoms, nor should they rely on taking the matter into their own hands and solving it without consulting a specialist. Most wrong decisions are made at this early stage of the disease, with serious consequences later on.

One typical situation is when the patients are already aware of something wrong, but put off consulting a doctor. If patients have bloody stools, they say it is hemorrhoids and probably postpone a medical consultation. They would probably attribute persistent cough to influenza, or a lot of work and stress. Patients would find excuses to put off the medical examination of these symptoms. This procrastination itself is enough for the disease to progress and be diagnosed in a worse stage. As a result, treatment options would be more limited, too. A key question for treatment is whether a patient is operable at the time of his or her diagnosis. If a patient, despite the complaints, visits the doctor too late, that means he or she may lose the chance to have the tumor surgically removed in time. Patients in this case deprive themselves of the possibility of being cancer-free after a successful surgery performed in the early stage of the disease.

The opposite is also true when a patient feels a loss of energy, has symptoms, and visits a doctor, but the real cause of the symptoms remains unknown. In such cases, patients should visit the doctor sometime later, in the hope that the repeated examinations do show a detectable (with imaging or by other means) anomaly which helps doctors in establishing a diagnosis. Unfortunately, patients rarely attend these repeated examinations, which again results in a delay in establishing a diagnosis, and treatments started much later.

One of the worst consequences is when a lump is identified but is diagnosed as a benign lesion and not treated in an adequate manner. A fundamental rule of establishing an oncological diagnosis is that a pathology record is needed to determine the qualities of the lesion, meaning that cells must be extracted from the tumor. In the case of breast and prostate tumors, a biopsy is a routine operation. During the biopsy, the tumor is punctured with a thin needle to extract cells from it. By examining the sample under a microscope, pathologists can decide whether or not malignant tumor cells are present. If no tumor cells are present, it can mean two things: (a) there are no tumor cells in the lump, or (b) the needle took a sample from a location where there were no tumor cells. A negative result is welcomed by everyone and the danger is considered to have passed, but the relief is only temporary because over time the symptoms can become more severe and the correct diagnosis (albeit belatedly) is finally made.

A typical example of this in older men is an increase in the levels of PSA, a tumor marker for prostate cancer. In these cases, it often occurs that even if the biopsy is performed several times, no tumor cells are found in the sample and thus the treatment cannot begin.

The difficulties of diagnostic tests

Establishing a diagnosis requires several tests, which in most cases confirm the existence of cancer. Yet, some unanswered questions may delay the beginning of proper treatment.

- The tumor is localized, but it cannot be decided whether it is a tumor or a metastasis.
- Imaging tests clearly show a structure of unknown nature.
- The tumor is located in a place that is hard or impossible to reach by biopsy.
- High tumor marker levels indicate the presence of a tumor, yet it cannot be located.
- The presence of tumor cells in the biopsy sample is clear, yet the origin and histology of the tumor cannot be determined.
- The result of a pathological examination is confusing; pathologists cannot decide whether the sample contains tumor cells or not.

In such cases, patients are subjected to several additional tests (physical examination, imaging tests, blood tests) to establish an accurate diagnosis, meaning further delays in starting conventional treatments. Everyone's primary goal is to ensure that the patient's chances of recovery and outlook are as strong as possible. To achieve this, the diagnosis must be correct to ensure that the patient receives the most appropriate therapy. Given the complexity of cancer, it can only improve the patient's chances if several professionals offer their opinion on the patient's condition. Unfortunately, the complexity of the disease also means that specialists often come to conflicting conclusions. One way of reducing uncertainty is to seek further opinions from other specialists, but there is an optimum point in this process when seeking further specialist opinions becomes an obstacle in starting treatment. We have to face the fact that absolute answers are rarely given, but decisions have to be made even when pros and cons are present simultaneously.

THE APPLICATION OF DEUTERIUM DEPLETION IN BORDERLINE CASES WHEN IT IS REASONABLE TO SUSPECT CANCER, BUT NO DIAGNOSIS AND TREATMENT IS AVAILABLE YET

In the transitional period, when it still cannot be decided whether the disease is malignant or whether the complaints are caused by something else, patients are recommended to consume DDW following the P/D (Prior to Diagnosis) Protocol in case of complaints. In this case, DDW can inhibit tumor growth and induce tumor cell death before the conventional treatments are started, contributing significantly to the success of subsequent conventional treatments. Based on the experience of the past years, it's safe to say that the consumption of DDW has no harmful effects in the examined concentration range (25–125 ppm deuterium). (However, it does not imply that the treatment should be started immediately with DDW at a concentration of 25 or 45 ppm deuterium.) These results guarantee that DDW can be used safely without any fear of toxic effects during its consumption. More importantly, no cases were suggesting that DDW would stimulate tumor growth.

The mere fact that something is harmless does not constitute a strong enough foundation to recommend its use in cases when a cancer-related problem is the root cause of the symptoms. However, not only the safe use but also the effectiveness of DDW is considered verified. Therefore, a deuterium concentration of 105 or 85 ppm is recommended until an accurate diagnosis is made. It must be considered, however, that the consumption of DDW may affect any potential checkups as discussed earlier.

In fortunate cases, symptoms are no longer present by the end of the four months of consuming DDW 105 or DDW 85, with further examinations unnecessary. However, it is important to be aware that there may be several reasons why the symptoms disappear:

a) The complaints would have been resolved even without consuming DDW 105 or DDW 85,

b) the root cause of the problem was not cancer-related, however, consuming DDW caused the symptoms to disappear,

c) a cancer-related issue is the root cause of the symptoms. In such cases, where the disease has not been identified until the symptoms have resolved, the patient should be referred back for regular checkups.

If the tests do confirm the presence of a tumor, the use of the P/D Protocol also means that treatment of the tumor has started before a definitive diagnosis has been made.

Establishing a diagnosis and communicating it to the patient

By the time a diagnosis is made, a patient has usually visited different clinics for weeks or even months, undergoing different tests. The long wait, the stories they have heard, the uncertainty, and the doubts about the future are a huge burden for patients and their families. At the end of the lengthy process, patients are notified in the form of a brief, written description or (in more fortunate cases) directly from their physician that specific cancer has been found and in which stage the disease is. These two pieces of information provide the doctor with a sound basis for judging the chances of a cure for the patient.

The statistics on survival for a given tumor type are based on data from hundreds of thousands of patients and over decades of observations. A doctor can rely on these statistics and make a prognosis for a patient. If the prognosis is good, it gives the patient strength and faith, and they are happy to receive encouragement. But if the prognosis is poor, it just reinforces a feeling of hopelessness in them. The major shortcoming of statistical analysis of cancer, which is also true in other cases, is that it is at the extremes, making it difficult to see clearly. In the context of cancer, neither patients nor doctors know whether the lower or the upper extreme of the typical statistical life expectancy is expected in a specific case. While the most common result is a value between the two extremes, there may be an up to ten times difference between them. In terms of doctor-patient communication, it is important to give a positive prognosis, but it is also important to stress that unjustified, excessive optimism can also be problematic. In the case of a tumor that has been detected and removed in time, doctors often reassure patients that they can consider themselves cured completely, that no further treatment is necessary and that they should not be bothered with the disease for the rest of their lives. It is a frequently heard opinion after surgery that the tumor has been completely removed, yet the disease does recur years later.

It can be misleading to inform the patient after successful surgery and removal of the tumor along with some intact tissues that they can consider themselves to be completely and finally cured. It is also misleading to claim that the tumor was discovered in time. Of course, the above circumstances are not at all irrelevant to the chances of being cured. The reality, however, is that tumors may be detected either early or late, but never in time.

Malignant tumors are characterized by the ability of cells or groups of cells to detach. These cells may then enter the lymph and bloodstream and travel to distant parts of the body. Tumor cells may have already migrated from the primary tumor even before it is diagnosed. Fortunately (and partly due to the immune system), a substantial part of these migrating cells are never capable of attaching to the surrounding tissues and forming a new tumor. The reason why cancer often recurs after successful surgery is that these cells sometimes manage to get out of the immune system's control and circulate in the body. Therefore, although there is reason for optimism after successful surgery, it is unfortunately not possible to declare that all the tumor cells have been removed from the patient.

The complementary use of deuterium depletion as part of the aftercare following a successful surgery can be an adequate response to this challenge and can significantly contribute to preventing the disease from recurring.

Of course, many components contribute to the success of therapy. Positivity and an optimistic mindset play an important role and therefore need to be reinforced consistently. The integration of deuterium depletion into a treatment regimen offers an opportunity to monitor the patient's condition with more optimism. Adding deuterium depletion to the therapeutic toolkit provides a solid reason for optimism.

In Chapter Fourteen, *"Advice on establishing the dosage,"* we discuss in detail the use of DDW depending on the disease, stage, and conventional treatments used.

Planning a treatment after diagnosis

The treatment of patients is nowadays decided jointly by a cross-disciplinary team. Experience gained from oncology treatments is constantly reviewed by physicians and they decide about using internationally recognized treatment protocols. The discipline of oncology has a right to be proud of the results achieved. However, this is overshadowed by the fact that despite all efforts, thinking, knowledge, experience, and financial investment, cancer is still responsible for the death of 8.8 million people every year. This raises the question of whether doctors are fully exploring the possibilities of a cure, or whether they merely use the toolkits offered by the pharmaceutical industry. A typical example of this is when a patient asks their doctor about dietary restrictions, and the simple answer is to eat whatever just feels good. Given the metabolism of tumor cells, a responsible professional should never give such advice. The least a doctor could do is recommend the restriction of carbohydrate intake.

Appalling international death statistics alone raises the question of whether this global challenge is addressed in the right way.

Deuterium depletion may be integrated into treatment regimens in two phases. Currently, we are in the first phase when the anti-cancer effects of DDW are recognized only in veterinary medicine, despite pre-clinical, prospective, and retrospective human clinical experience clearly confirming the anti-cancer effects of deuterium depletion. In this phase, the discipline of oncology is not able to take advantage of the benefits of DDW, which substantially increases patients' chances for survival. The second phase begins when the anti-cancer effects of the procedure are accepted in the field of human clinical treatment. In this phase, DDW is considered not only as a registered active substance but also as a human drug when it comes to creating a treatment plan. We can reasonably expect to transition into this period shortly. Until then, we continue to publish scientific papers and impart scientific knowledge in professional forums to provide a solid basis for the future.

Difficulties in planning conventional treatments

Thanks to extensive research over the past decades, the treatment options for curing cancer are ample. Despite a large number of treatment options, the majority of patients are treated with the following procedures and combinations: surgery, chemotherapy, radiotherapy, hormone therapy, and immunotherapy. Based on the available scientific and clinical evidence, it is reasonable to expect that the application of deuterium depletion will lead to a breakthrough and will redefine treatment protocols in the future. Yet, it is important to emphasize that the effectiveness of oncotherapy can improve significantly if DDW is integrated into the current treatment regimen and not solely used instead.

One of the major goals of writing this book was to collect and provide a summary of the scientific knowledge and guiding principles to help healthcare professionals (primarily oncologists) consider deuterium depletion as an addition to conventional treatment options on a solid scientific basis.

The book describes how deuterium depletion can enhance the effectiveness of surgery, chemotherapy, radiotherapy, and hormone therapy, and also, to what extent conventional treatments can enhance the effectiveness of DDW. (There is currently no data available on the interaction between DDW and immunotherapy.)

There are several factors and uncertainties that may make the planning of a patient's treatment difficult (even though there exist well-defined protocols). Some examples of the main criteria to be considered during the entire period of the treatment:

(a) In the case of a small, well-localized breast tumor, it becomes clear only during surgery that the surgical area must be extended. This decision depends on the findings from the examination of the sentinel lymph node.

(b) A pathology test of the tumor dictates which medicine should be used for the treatment (for example, considering the hormone status).

(c) The findings from the pathology test help decide – and only usually weeks after the surgery – whether surgeons removed the tumor with the surrounding intact tissue successfully. If not, then repeated surgery and the removal of even more intact tissue is necessary.

(d) An additional difficulty arises from the fact that surgeons plan the surgery based on imaging tests. During surgical exploration, however, it may become evident that they are facing a different situation than expected. One possible outcome is that the surgery is completed and the wound is closed without removing the tumor, or a more complicated surgery is needed than expected.

(e) In very critical situations, surgeons have to decide on the spot whether it is necessary to remove larger areas of tissue and choose a more drastic method that substantially affects the patient's quality of life. One example is a patient with rectal cancer. It becomes clear during the surgery whether a stoma is needed, or the anus would be retained.

(f) Inoperable tumors may require pre-treatment with radiotherapy and/or chemotherapy. While there are several chemotherapy treatments, it cannot be predicted with absolute certainty whether the drug of choice is effective or not. There is also a lot of uncertainty about how patients would tolerate severe side effects.

(g) During treatment, some complications may arise that cause deviation from the original treatment plan. During chemotherapy treatment, blood counts often deteriorate to such an extent that treatment has to be suspended, and during radiotherapy, the treatment may have to be interrupted due to severe skin irritation of the affected area.

The use of deuterium depletion can significantly reduce the side effects of conventional treatments and improve their tolerability. The results are monitored and analyzed throughout the treatment period for changes and unforeseen events, and if necessary, the therapy is adapted to accommodate these changes. This type of continuous monitoring should be maintained throughout the application of DDW. It is recommended that the use of deuterium depletion should be aligned with conventional treatments and if these change, a change in the DDW dosage may be necessary. In addition, it is recommended to monitor changes in the patient's condition, to adjust the dosage of DDW to the results of control tests, and to change the deuterium concentration after prolonged use of a given DDW concentration, or to suspend DDW consumption. Attention should also be paid to the length of breaks in DDW consumption, which is discussed in more detail later.

CHAPTER EIGHT

Additional Use of Deuterium Depletion Alongside Oncotherapies

The discovery of a submolecular regulatory system in cells based on the changes in the ratio of deuterium to hydrogen opens up new possibilities for more effective treatment of patients. However, this does not imply that the current therapeutic methods should be abandoned as deuterium depletion is most effective in the treatment of patients when properly combined with conventional treatment methods. Synchronizing oncological procedures with the tools of submolecular medicine takes a considerable amount of time and will change a lot in the future. For the time being, we can rely on the knowledge, results, and clinical experience gained and published by a key player in the field, HYD LLC for Cancer Research and Drug Development. The next part of the book describes deuterium depletion as a complementary treatment method and the optimal combination of it with conventional treatments. Discussed are the possibilities for combining the four most common conventional treatments (surgery, chemotherapy, hormone therapy, radiotherapy) with deuterium depletion. Recommendations are made on how to integrate DDW into the treatment regimen and how to properly time it.

THE TIMING OF DEUTERIUM DEPLETION AND CONVENTIONAL TREATMENTS ARE DIFFERENT

The highest degree of efficacy may be achieved through the proper synchronization of DDW and conventional treatments. In some cases, however, the consumption of DDW does not take place simultaneously with conventional treatments. This may happen in the following cases:

(a) The conventional treatments have not yet started,

(b) patients have not consented to the offered treatment,

(c) the conventional treatment options have been exhausted,

(d) the treatments have already been completed.

The conventional treatment according to the protocol has not started yet

Chronic lymphocytic leukemia (CLL)

CLL (chronic lymphocytic leukemia) is a slowly progressing disease that mostly occurs in old age. Symptoms of the disease include an elevated white blood cell count exceeding 10,000 and enlarged lymph nodes. For this type of hematopoietic disease, according to the treatment protocol, it is not necessary to start medication treatment after diagnosis. Rather, the patient's survival is best secured by starting chemotherapy as late as possible after diagnosis. Therefore, if there are no other contraindications, it is recommended to wait until the white blood cell count reaches ten times the normal level, that is, 100,000.

A 41-year-old male patient was diagnosed with CLL in February 2006. In addition to a white blood cell count of 16,000, he had a significant enlargement of the lymph nodes on the neck, a spleen twice the normal size, and an ultrasound scan showed a nine-centimeter lymph node conglomerate in the abdominal region. The patient started consuming DDW immediately after the diagnosis. Three months later, the white blood cell count decreased to below 10,000 in the normal range and the size of the lymph node also changed favorably. The patient first consumed DDW without interruption for more than three years, during which time the size of the spleen went back to normal. The abdominal lymph node conglomerate broke down into smaller lymph nodes and the enlarged neck lymph node almost completely regressed. Later, when the DDW treatment was interrupted, a minimal progression could be detected. Even this minimal progression could be reversed with a DDW treatment. During the past thirteen years, the patient has consumed DDW twelve times in several-month intervals after the first treatment. To this day, chemotherapy has not been required.

The case illustrates the anti-cancer effects of DDW, as there was a clear relationship between DDW intake and the reduction in the size of lymph nodes, and also, between the interruption of DDW treatment and the consequent progression of the disease. For patients with CLL, the use of DDW may postpone the start of conventional treatments and may even render their use unnecessary.

The patient has not consented to conventional treatment as prescribed by the protocol

Prostate cancer

A 68-year-old patient was diagnosed with prostate cancer in late October 2009, with elevated PSA levels (8.7 ng/mL). Following his recent diagnosis, the patient did not accept the hormone therapy offered due to the resulting impotence. He started consuming DDW a month after the diagnosis, with PSA levels dropping to 6.3 ng/mL within a month. Subsequently, 5.28 ng/mL was detected in January 2010 and 5.15 ng/mL in March 2010. In the first phase, the patient consumed DDW for nine months without interruption. PSA levels fluctuated between 5.0 and 8.0 within this period. Before the discontinuation of the DDW treatment, the PSA level was 4.68 ng/mL. The patient started a new DDW course after a two-month break, in November 2010. Following the resume of DDW, the patient repeated courses with a four to five-month duration each, with PSA levels eventually stabilizing at 4–5 ng/mL. A spike was detected in the summer of 2010 after the patient had spent a few days at a thermal bath. Following this event, PSA levels increased to 24 ng/mL, only to start dropping again to 5 ng/mL within two months (see Fig. 19). In October 2013, the previously one-centimeter large tumor could not be detected anymore with imaging tests. The patient took five to six months of pause between each DDW course (lasting three to four months) in the following four years. Due to the stable low PSA levels, the patient interrupted the consumption of DDW for eleven months. Nine months after this, the elevation of PSA levels indicated that the prolonged interruption of DDW consumption was favorable for tumor cells. Nine years after the initial diagnosis and six years after the result that the tumor could no longer be detected, an MRI scan detected a prostate tumor.

The patient resumed the consumption of DDW. At the time of writing, he has had no symptoms or complaints. Figure 19 shows the evolution of PSA levels. The only spike (exceeding 20 ng/mL) in PSA levels was detected a few days after the patient spent a few days in a thermal bath. (The case is also described in the chapter, *"Dilemmas of histological sampling."*)

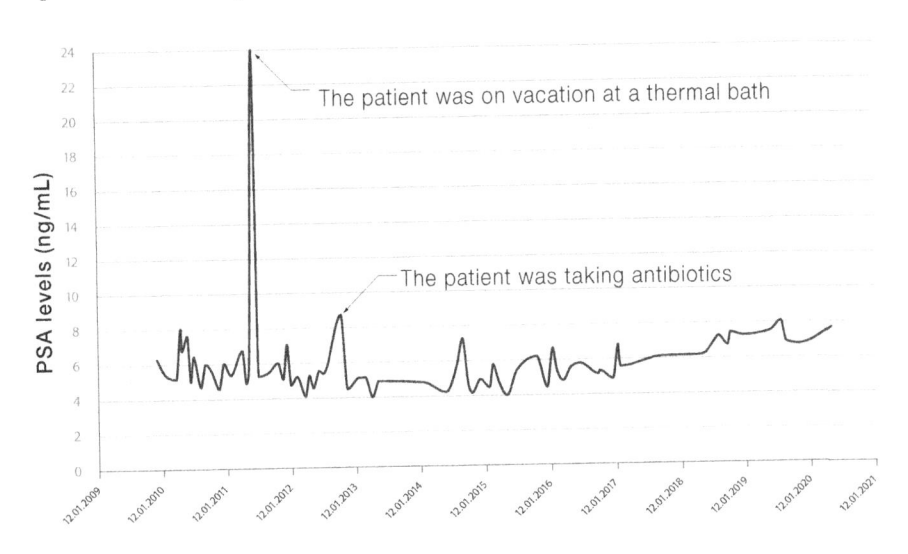

FIGURE 19
The evolution of PSA levels (in ng/mL) in patients who were diagnosed with prostate cancer and who did not receive conventional treatments

Experience has shown that the proper timing of DDW consumption and interruptions helps maintain a condition in which the disease does not progress, and the patient maintains a good quality of life. However, it is difficult to establish a general rule that is valid in every situation. Sometimes the disease progresses even when small interruptions are scheduled. Some other times, patients remained in remission even after years of interrupting the consumption of DDW. A guideline might be if a schedule (DDW consumption – interruption) proves to be effective. This schedule should then be kept and modified only after very carefully. Established dosage protocols for different situations provide some clues for such modifications.

Breast cancer

A 49-year-old female patient was diagnosed with breast cancer in December 1995. The patient did not consent to conventional treatment. She started consuming DDW in February 1996. Within a few months, the tumor started to recede from the nipple. Within seven months, the size of the tumor reduced from the former 20 × 30 mm to 20 × 23 mm. By February 1998, the size of the tumor reduced to 10 mm, with frequent oozing. In 2001, the tumor moved close to the skin surface and was soft when touched. (This would have been the ideal time for surgery. However, the patient repeatedly declined it.) Until 2003, the patient consumed DDW only and then significantly reduced the daily DDW intake in the following four years. She reported having lost ten kilograms in two years. In late 2007 she reported a persistent cough, however, the size of the tumor did not change. Ultimately, the patient died in the spring of 2008, twelve years after being diagnosed with breast cancer. In all likelihood, she developed distant metastases. However, since she did not return for regular checkups, we don't have any information on metastases.

Another patient was diagnosed in 2002 with breast cancer. The patient rejected the offered treatment and started using deuterium depletion. In the first few years, she consumed DDW regularly and exclusively. A reduction in the size of the tumor was detected, and the tumor started receding from the nipple, a clear indication of regression. Only much later, eight years after the diagnosis, progression was detected as a result of irregular consumption of DDW. One year later, the patient was administered hormone therapy and started to consume DDW again. The size of the tumor decreased from 33 mm to 28 mm in just a few months. This condition was maintained for a year before the disease progressed despite conventional treatments. The last piece of information on the patient was received eleven years after diagnosis, by which time she had already developed distant metastases.

The above two examples indicate the anti-cancer effects of DDW, but they are also a warning sign that patients should not turn down surgery when recommended. Surgery should not be rejected even when the size of the tumor makes it necessary to perform a mammectomy. Deuterium depletion helps achieve significant regression and keep the disease under control. If a completely cancer-free condition cannot be achieved, then suspending the consumption of

DDW or the irregular consumption of DDW causes the persisting tumor cells to engage in cell division and the tumor starts growing again. If patients had started consuming DDW immediately after the surgery, even better results than the ones above could have been achieved. The cases of patients who followed the protocol prove the above point.

Lung cancer

A 76-year-old female patient was diagnosed in March 2016 with lung cancer. Thoracentesis was performed to drain 1.7 liters of fluid from the lungs. The patient refused chemotherapy and decided to use deuterium depletion one week after the diagnosis. Two weeks later, 1.5 liters of fluid was drained from the lungs, followed by another liter two weeks afterward. Two months later, 1.2 liters, and half a year later, only 0.4 liters of fluid was drained. From October 2016 to the time of writing (2020) no further thoracentesis was performed. The persistent cough of the patient stopped. She is physically active and is in good general condition three years after the discovery of the tumor. A CT scan from April 2018 showed no tumors on the right side, as opposed to the CT scan from July 2017.

Conventional treatment options for lung cancer patients are limited. The effectiveness of these therapies is also below that of the tools used in the treatment of breast cancer. Yet, the calculated median survival time (MST) for 300 lung cancer patients using deuterium depletion was six times longer (48 months) that of the patients receiving conventional treatment. This is considered as a result of the combined use of DDW and conventional treatments.

Conventional treatment options have been exhausted

Stomach cancer

A 63-year-old male patient was diagnosed in January 2016 with stomach cancer. As a result of conventional treatment, a CT scan in April confirmed regression. Subsequently, the disease progressed, with the patient having dark-colored stools. A follow-up checkup in October revealed that the tumor occupied the entire stomach, and metastasis was detected in the liver. The patient started consuming DDW in late August. By the end of the first month, he experienced extreme fatigue and nosebleed. Two months into the consumption of DDW, yellow and white mucoid plaques appeared in the normal-colored stools. The patient was able to consume more foods and within two months, he gained ten kilograms of weight. In April, the liver metastases were no longer detectable. The size of the stomach tumor was also reduced. Pain and other complaints disappeared. Subsequently, the patient discontinued the use of DDW, and his condition rapidly started to deteriorate. He then resumed DDW consumption and the symptoms became milder, but the patient died two months later, in July 2017.

Liver cancer

A male patient was diagnosed in 1985 with a malignant tumor originating from the connective tissue of the liver. During the nine years before DDW treatment, several surgeries to reduce the size of the tumor and vascular surgery to stop the blood flow to the tumor were performed. In May 1994, due to tumor-related stomach bleeding, a partial gastrectomy was performed and bowels were surgically connected with the stomach. The patient developed obstructive jaundice and loss of consciousness due to repeated episodes of hypoglycemia and was admitted to the hospital in October 1994 due to his drastically deteriorating condition with extremely high bilirubin levels. An ultrasound scan in November revealed a 17 × 21 cm tumor. Following the consumption of DDW, the patient developed a healthy appetite, gained 3–5 kg weight, and was able to walk again. He was discharged from the hospital in late November. The patient's jaundice measurably improved and the elevated enzyme levels decreased. In February 1995, an ultrasound scan confirmed only a reduced, 16 × 14 cm tumor. One year after the start of DDW treatment, the jaundice was barely detectable. The normal flow of bile was confirmed by the color of the patient's stools going back from clay-colored to normal. A significant deterioration and urinary tract obstruction occurred in late November and early December. The patient was then no longer available to consume DDW and died in December 1995. The significant improvement in the patient's condition starting in November 1994 was maintained only for a year and can only be attributed to the use of DDW (the patient received no other treatment during that period).

Experience suggests that a minimum of 2–3 months of life expectancy is necessary at the beginning of the DDW consumption for deuterium depletion to take effect (this does not imply a complete recovery from the disease). In this stage, treatment options are very limited, and DDW should only be used as a last resort. We have every reason to be hopeful, as when our knowledge and experience grow, we will become capable of reversing malignant processes even in a late stage, as cases from veterinary medicine suggest. Such cases are discussed in detail in Chapter Fourteen, section *"Special advice on establishing the dosage."*

Conventional treatments are completed

Conventional treatments ensure achieving a cancer-free condition for a given cancer and stage. Conventional treatments are followed by a period of passive waiting, hoping that check-ups would not confirm the recurrence of the disease. The use of deuterium depletion as part of the aftercare offers a tool to maintain a cancer-free condition for a complete and ultimate recovery.

Results from this group of patients were introduced in the chapter, *"Deuterium depletion helps prevent and protect against the recurrence of cancer."*

THE COMBINED USE OF SURGERY AND DEUTERIUM DEPLETION

One of the keys to successfully treating cancer is surgically removing the tumor. Only a minority of oncology cases are classified as operable at the time of diagnosis. In some cases, the tumor may become operable due to radiotherapy, chemotherapy, and/or hormone therapy. A primary goal of surgery is to remove the tumor and its metastases, as it's the only way that patients can become cancer-free. If only one out of three tumors are removed, then patients are exposed to surgical stress and risks without actually improving their chances for survival. Such surgical procedures may only be justified if there are other, life-threatening conditions, such as intestinal obstruction.

The preoperative use of deuterium depletion

Over two decades of veterinary experience suggests that consuming DDW before surgery may improve operability. For recently diagnosed and operable cancers, it is recommended to start deuterium depletion at least two to three weeks before surgery. Postponing an otherwise pending surgery may be an option when DDW is used.

As a result of deuterium depletion, tumor cells start to necrotize in and around the tumor, making the tumor more mobile within a few weeks. This is because the tumor tissues infiltrating the surrounding tissues die and recede as a result of necrosis. A few weeks of DDW consumption before surgery reduce the size of the tumor, within which time any inflammation is reduced or even eliminated. As a result of DDW consumption, tumor-affected and healthy tissues are easier to separate, contributing to the success of the surgery.

A reason against putting off surgery is that removing the tumor as soon as possible improves the patient's chances for survival. However, since a detected tumor may have been growing in the patient for up to four to five years (208 to 260 weeks), potentially delaying the surgery for two to three weeks extends the tumor's presence in the body by only 1%.

Picking the ideal time for a surgery

The optimal time for surgery is when a tumor can be removed with the least possible damage, with the highest degree of safety, leaving the surrounding tissues intact. If a patient consumes DDW, it may reduce the size of the tumor. This also raises the question of when the surgery should take place. As a result of deuterium depletion, ideal conditions may be present a few weeks later.

Choosing an ideal date for the surgery is of crucial importance for elderly patients. Another important factor is whether there are underlying chronic conditions and whether they are treated. If the risk of surgery is too high and deuterium depletion helps achieve regression, postponing the surgery may be an option.

For a given DDW level, it is not recommended to postpone surgery by more than two to three months. If the deuterium concentration used is reduced and the regression continues, then postponing surgery may be subject to consideration.

If a patient was diagnosed in an early stage of the disease and there is still sufficient time to consume DDW, the deuterium concentrations of 105 or 85 ppm are recommended (*C/C/Op Protocol*).

For operable tumors, consuming DDW for a few weeks with concentrations 105 or 85 ppm may cause the reduction or disappearance of tumor infiltration in the surrounding tissues. The tumor becomes mobile and is easier to locate during surgery. Chances for removing the tumor while leaving the surrounding tissues intact improve. For deuterium levels between 105 and 85 ppm, the decision of whether to postpone a surgery is also influenced by the age of the patient, the amount of time left until the surgery, and the histology and size of the tumor. For a tumor of well-defined location and with a good prognosis, patients with average body weight may consume DDW with a concentration of 105 ppm without interruption. If the patient has been diagnosed with aggressive cancer with a bad prognosis and uncertain operability, it is recommended to start a DDW treatment with lower deuterium levels (85 ppm).

Head and neck cancers

Head and neck cancers form one of the most critical groups of cancers due to the potential need for resection and a loss of quality of life. Cited here is the example of a patient who was diagnosed with tongue cancer and was operated on twice until 1992. The date of the third surgery was also decided, when in December 1992 the patient started consuming DDW. Four months later, the surgery was postponed due to a regression. The correctness of this decision was later confirmed by the fact that no tumor cells were detected in a biopsy sample.

Bladder cancer

A patient underwent several surgeries for bladder cancer. However, the disease recurred despite the repeated surgeries. The patient consumed DDW before an upcoming surgery. During the surgery, the tumor was easily separated from the inner lining of the bladder. The patient has not relapsed thanks to the consumption of DDW and the disease has not recurred since.

Breast cancer

In many cases, the consumption of DDW reduced tumor size in breast cancer and increased the operability of cancer.

In another case, a pre-operative Herceptin treatment was administered to a patient to improve operability. Combining DDW with Herceptin resulted in a complete regression, which was confirmed by a pathological examination following the surgery. No tumor cells were found in the surgically removed tissues.

The postoperative use of deuterium depletion

For a smaller group of patients, surgery may be performed without pre-treatment. For patients operated on without pre-treatment, based on the appearance of the surgical area, results from pathological examinations, and molecular biology tests, a team of oncologists may decide that there is no need for aftercare. Deciding that there is no need for aftercare may be justified if chemotherapy, radiotherapy, or hormone therapy are aftercare options. The above treatments may have harmful side effects and there is a minimal chance that the disease would recur. Conversely, the post-operative use of deuterium depletion does not have any negative consequences but may help to reduce the risk of recurrence. In all cases (even when the current protocols do not require it), following the *C/R/1 Protocol* after surgery is recommended.

We have previously discussed that during the four to five-year-long period of cancer formation, tumor cells break off of the primary tumor and drift to distant parts of the body through the bloodstream or lymph. Therefore, tumors may reappear even after successful surgery and the disease may recur. The consumption of DDW may significantly decrease the risk of recurrence.

THE COMBINED USE OF DEUTERIUM DEPLETION AND CHEMOTHERAPY

The two main approaches to chemotherapy have evolved under the influence of pharmaceutical development in the past decades. Until the 1970s and 1980s and prior to the advancements in genetics, molecular biology sought to find molecules with anti-cancer effects by testing thousands of molecules (in cell cultures and animal testing) for medicinal purposes. While undoubtedly achieving some success (to this day, many such products are used in oncotherapy), it became clear that this approach would not solve the treatment of cancer patients. The majority of chemotherapy treatments do have severe side effects, and cancer progresses after a temporary improvement and a significant proportion of patients die of the disease.

The results of genetics and molecular biology research and the use of targeted treatments brought significant improvements. Development now is being conducted with more awareness. Inhibitory molecules are being designed with the protein structure in mind. Drugs target specific processes, significantly lowering toxicity.

Over the past nearly thirty years, patients using deuterium depletion have been treated with both products, so the results are also the combined results of both treatments, and the effects cannot be attributed solely to deuterium depletion. However, with our broadening knowledge, some conventional treatments may be excluded from current therapies. In many cases, these therapies often provide minimal benefits with severe side effects.

General advice for the complementary use of deuterium depletion alongside chemotherapy

There are a large number of drugs available for the treatment of cancer, targeting different processes. It is not our intention to give a detailed description of these drugs, but rather to provide general principles for the effective combination of chemotherapeutic treatments and deuterium depletion that can be followed to achieve the best synergistic effect.

Of course, the primary goal of all oncological treatments is to destroy tumor cells in the body, but there are significant and fundamental differences in the mechanism of action between chemotherapy and deuterium depletion. It is therefore important that the combination of the two types of treatment is based on a thorough understanding of the methods.

One fundamental difference is that cytotoxic chemotherapy treatments have severe side effects. Patients may only be administered a certain number of treatments, usually with interruptions between them. Until the first few treatments, it is impossible to assess the effectiveness of a therapy or the side effects. For the above reasons, chemotherapy treatments are generally provided in a hospital setting.

Neither pre-clinical trials (cytotoxicity tests, animal testing), nor decades of using deuterium depletion have shown any unexpected or harmful side effects. This creates a new situation and opens up new perspectives in oncology, as there is no time limit for the use of DDW. An example supporting this argument is the case of a patient who has been diagnosed with several large melanoma metastases in the liver. At the time of writing (summer of 2020), the patient has been consuming DDW for twenty-six years and has enjoyed good general health. To the best of our knowledge, deuterium depletion does not have any toxic side effects. However, there is no data available on whether extremely low deuterium levels (lower than 10 ppm) can cause any undesired effects.

A further difference in the mechanism of action is that chemotherapy targets only a single process, whereas the lower deuterium concentration induced by DDW also reprograms cellular metabolism and genetic processes, improving mitochondrial function and thus cellular energy balance. As a combination of these effects, aggressive tumors may behave differently. The pathology of tumor cells may resemble that of healthy cells. (This question is also discussed in the chapter *"Dilemmas of tissue sampling."*)

To give a summary of the above, deuterium depletion may increase the effectiveness of chemotherapy and mitigate its toxicity. It protects hematopoietic organs and thus contributes to keeping the planned schedule of conventional treatments.

The use of deuterium depletion following chemotherapy

Integrating deuterium depletion into a conventional treatment regimen does not call for immediate use (following diagnosis) of DDW. In cases where the disease is kept under control with chemotherapy, DDW may be integrated into a later phase of treatment. Examples include childhood leukemia (AML, ALL), testicular cancer, Hodgkin's lymphoma, and non-Hodgkin's lymphoma. In such cases, conventional treatments have a high chance of bringing a rapid and significant improvement, or even complete remission. In such cases, deuterium depletion should be used in the last phase of treatment, or when the check-ups confirm a regression.

Also, deuterium depletion may be used later in cases where the disease has a good prognosis (e.g., starting chemotherapy as an adjuvant after a surgery). Such a case is where a breast tumor smaller than 2 cm is removed, no lymph nodes are affected, and the hormone status is positive, or also when colon cancer is operated on in an early stage.

Oncological treatments ensure that most patients are macroscopically rendered cancer-free (as shown by imaging tests). Yet, a flaw of oncological treatments is that cancer recurs in a significant proportion of patients. Sooner or later, metastases appear, and at this point, life expectancy drastically decreases. This shows that while conventional treatments do reduce the number of tumor cells, they cannot destroy all of them. A delayed integration of DDW into the treatment regimen may extend the period spent in a cancer-free condition and maintain it. Integrating deuterium depletion into the treatment regimen at this point has the benefit that at the end of the conventional treatments, the number of tumor cells is the lowest. In this situation, DDW may help achieve a completely cancer-free condition. Currently, there is no test method available that would provide solid proof to say that tumor cells (with the potential to develop into cancer) can no longer be found in a macroscopically cancer-free patient. Conversely, everyone can develop a tumor at any moment, as the human body creates billions of cells every day. Therefore, our advice is that those who have already been diagnosed with cancer and became cancer-free as a result of the treatment(s), should consume DDW according to the C/R/1 Protocol or the C/R/2 Protocol.

The simultaneous use of chemotherapy and deuterium depletion

Oncological treatments in several cases cannot be expected to result in complete regression. In these cases, deuterium depletion may be used simultaneously with other treatments. However, conventional therapies may impair the effectiveness of deuterium depletion as they can deteriorate blood test results and weaken the immune system. The effectiveness of deuterium depletion may be adversely influenced if vomiting and diarrhea occur as a result of chemotherapy, or if patients receive an infusion with normal deuterium levels. While DDW may enhance the anti-cancer effects of chemotherapy, the reverse is not always true. We, therefore, recommend rescheduling increasing the dosage of DDW to the period after the chemotherapy treatment when using chemotherapy and deuterium depletion simultaneously. In the presence of chemotherapy, DDW 105 ppm or DDW 85 ppm may be sufficient to support conventional treatment and mitigate toxic effects (*1/C/C/Chem Protocol*).

Deuterium depletion is started after or at the same time as chemotherapy treatment producing partial results

In a significant proportion of cases, a cancer-free condition is not achieved despite the treatments. In this situation, one option is to temporarily suspend treatment in the hope that the disease will progress slowly or that the patient's general condition will improve as the toxic side effects disappear and treatment can be resumed later. Another possibility is to continue treatment with other chemotherapeutic agents that are also harmful to the body. There may come a point, however, when there are no therapeutic benefits in experimenting with other treatments.

In this situation, we recommend the application of DDW according to the *1/C/C/C/Chem Protocol*, starting with DDW containing 85 ppm deuterium. The only restriction is that the concentration should not be lowered every two to three months, but every one to two months.

THE COMBINED USE OF DEUTERIUM DEPLETION AND HORMONE THERAPY

Hormone therapy is mainly used for gynecological, breast, and prostate tumors. The therapeutic benefit of this intervention is based on the fact that hormones circulating in the blood often stimulate the proliferation of tumor cells, so lowering hormone levels or blocking hormones binding to tumor cells can significantly inhibit tumor cell division and slow tumor growth. Hormone therapies contribute significantly to the successful treatment of breast, ovarian, and prostate cancers, which are reflected in the relatively long-life expectancy of this patient population. A drawback of this treatment is that years of hormone antagonist drugs can lead to a predominance of cells within the tumor that is no longer affected by the hormone preparations, and the tumor can then start to grow again. This is the consequence of the fact that, according to commonly used modern protocols, patients receive hormone treatment continuously for years, even if they are in remission. If the disease progresses again, the patient switches to another hormonal drug or chemotherapy with a similar mechanism of action, with little or no significant improvement. The use of deuterium depletion alongside hormonal treatments could radically change the currently used protocols.

General guidelines for the combined use of deuterium depletion and hormone therapy

The right combination of hormone therapy and deuterium depletion can prevent the development of hormone resistance. The treatment time may be increased by several times, making a complete recovery a realistic and achievable goal. Two guidelines should be observed to achieve this: (1) taking advantage of the synergistic effects of combining deuterium depletion with hormone treatments, (2) alternating hormone treatments with DDW (recommended if a patient is in remission). The simultaneous use of the two treatments is recommended if a recently diagnosed patient has high tumor marker levels and is not operable. In fortunate cases, combining deuterium depletion with hormone therapy helps achieve remission, after which it is possible to switch to alternating the two treatments. Alternating the two treatments is recommended when a patient has already undergone successful surgery, or if there has been a significant regression thanks to hormone treatments.

When establishing the dosage for patients who receive deuterium depletion and hormone treatments, it should be considered that none of the preparations should be used for so long that it allows patients to develop resistance and not respond to the treatment. The so-called intermittent treatment is accepted by professionals for the treatment of prostate cancer. During an intermittent treatment, hormone therapy is temporarily discontinued while PSA levels are low. By continuously monitoring the patient's PSA levels, hormone antagonist medication is reintroduced only when PSA levels start to rise again, as this will ideally bring the tumor marker back down. The concomitant use of DDW may significantly postpone the time of the next hormone therapy, or even render it unnecessary. DDW courses of a few months (and repetition thereof following a correct schedule) may keep tumor marker levels at a permanently low level.

A major difference between chemotherapy and hormone therapy is that while the former can only be given in limited numbers due to severe side effects, hormone therapy can be administered for years. This is because hormone therapy works by way of the body's naturally occurring molecules and physiological processes. Chemotherapeutic agents, on the other hand, are mostly non-natural molecules whose development is primarily focused on achieving strong cellular toxicity. The use of hormone preparations is characterized by a kind of reversibility. By reducing hormone levels, tumor growth is inhibited, but this can be accelerated if hormone levels are increased. Repeatedly reducing the levels allows for the inhibition of tumor growth.

The combination of deuterium depletion and hormone therapy is successful only if processes are maintained as reversible. This means that patients are given a certain treatment (hormones, DDW, or a combination of these) only until as long as necessary.

Hormone treatment prior to deuterium depletion

Similar to chemotherapy, deuterium depletion should not be immediately integrated into hormone therapy. Once patients have undergone successful surgery, are cancer-free and have low tumor marker levels, or are close to the normal range, then it is recommended to integrate DDW 2–3 months after the hormone therapy has started. Deuterium depletion should be combined as described in the chapter, "General guidelines for the combined use of deuterium depletion and hormone therapy," while observing the *C/R/1 Protocol*. The delayed use of deuterium deprivation is justified following similar considerations as for chemotherapy. Knowing that hormone therapy can be particularly effective in the initial phase of treatment, it is not necessary to exploit the potential of DDW already in this period. Deuterium depletion should be introduced after a few months of hormone therapy alone, or when conventional treatment no longer results in further improvement. In this case, DDW can fight the small population of tumor cells that may be present in the body, significantly increasing the chances of effectively reducing the number of tumor cells in the body.

Deuterium depletion and hormone therapy are used simultaneously

In a significant percentage of breast, ovarian, and prostate cancer cases, it is not possible to remove the tumor immediately and completely by surgery. If there are no distant metastases in the body, there is a possibility that the tumor can be removed by surgery with one kind of therapy or a combination of hormone therapy, chemotherapy, and radiotherapy. In these cases, DDW can be used concomitantly with conventional treatments, because every tool is needed to help achieve operability. The use of DDW in these cases is recommended according to the *C/C/Horm Protocol*.

If distant metastases have developed and operability is not achievable (such is the case of prostate or breast cancer patients with bone metastases), then in the first few months, patients should be administered a hormone treatment only. DDW should be included later in the treatment regime due to the above-mentioned reasons (at this point, deuterium depletion can be used to combat a smaller tumor mass).

Deuterium depletion should be started following a hormone therapy

One of the properties of hormone therapy is that it is administered for as long as the disease does not progress, even for years. When the patient's condition worsens despite the treatment, it is an indication that the cancer cells in the body have become resistant to the treatment. As mentioned earlier, this stage can be avoided or delayed if a patient in remission is given a break from hormone treatment and then uses deuterium depletion in the meantime. If the patient remains in remission while on DDW, DDW may also be discontinued later, as recommended in the *C/C/Horm Protocol*.

One common problem is when low tumor marker levels cause a patient to stop taking DDW, to resume only when the levels start to rise again. It is therefore recommended that after an interruption of a few months, patients should resume the consumption of DDW even if tumor marker levels are in the low range. After a long period of continuous DDW consumption of eight to ten months, the first break should not exceed two to three months. However, the length of the breaks can be increased later if tumor marker values allow.

SYNCHRONIZING DEUTERIUM DEPLETION AND RADIOTHERAPY

Radiotherapy is rarely used as a standalone treatment, but usually as a part of a complex treatment regimen. It may be administered in time before surgery, or after surgery to ensure that the affected area is tumor-free. Radiotherapy is frequently used before or after chemotherapy. It is used to improve patients' chances of survival, or (in a late stage of cancer) irradiating the affected area of the body as a palliative method to reduce pain. These cases are discussed in detail in the context of complex treatments.

The use of deuterium depletion is recommended during radiotherapy, due to the synergistic effects of the two treatments. Consuming 105 ppm DDW is recommended during, and four to five weeks after radiotherapy. Extending the use of deuterium depletion by four to five weeks is recommended if the radiotherapy treatment has lasted for five to six weeks, and patients have received the maximum recommended dose of 60 Gy. If patients were administered radiotherapy for a shorter period and with a lower dosage, then the duration of DDW consumption may also proportionally be shortened. Once the patient's aftercare is complete and a cancer-free condition is achieved, it is recommended to follow the *C/C/Radther Protocol*.

The combined use of radiotherapy and deuterium depletion multiplies the effectiveness of each one of the treatments. Local radiotherapy aims to destroy all tumor cells in the tissues surrounding the surgical area. These tumor cells may have detached from the tumor before or after surgery and migrated to distant parts of the body. The use of DDW during and after radiotherapy can significantly improve the chances that the areas surrounding the surgical site will indeed become cancer-free, and the effect of deuterium depletion will also be felt in distant parts of the body not affected by radiotherapy. The repeated use of deuterium depletion can take effect in the entire body in the years following the treatments, inhibiting the growth of tumor cells.

Fitting deuterium depletion to conventional treatments

In the majority of cancer cases, it is impossible to achieve a cancer-free condition and complete recovery with a single conventional treatment. It is much more common and typical that a combination of therapies is what is effective. Also, the combination of different therapies is effective only to temporarily halt or slow down the progression of cancer. Discussed below are some potential treatment combinations including DDW. The stages of the disease are divided into three major groups (as described earlier). A decisive factor in determining treatment strategies is whether it is possible to achieve a cancer-free condition with conventional treatments only (stage I). In such cases, in the first phase of treatment, DDW acts as a support to conventional treatments. However, it plays a crucial role in the second phase of treatments when it comes to maintaining a cancer-free condition. In stage II patients, the tumor impacts surrounding tissues and so the operability of the tumor is uncertain. Enlarged lymph nodes may be present. Achieving a cancer-free condition with conventional treatments is uncertain; therefore, deuterium depletion may help achieve it. For stage III patients with distant metastases, there should be two groups in terms of DDW application: (a) Stage III was already present at the time of diagnosis, (b) Stage III developed years into the treatment. A key aspect of an effective treatment strategy is to make good use of the resources available. Patients for whom using DDW is an option after the diagnosis should consider conventional treatments as the primary option. Using DDW should be planned in a way that enhances the effects of conventional treatment. Increasing the dosage should only be an option after regression, or after complete remission. Patients who consider using DDW after an extensive treatment (often as the only available option), should predominantly use DDW. This means that deuterium depletion should start with higher dosage (lower deuterium concentration), followed by a relatively rapid concentration decrease.

Using deuterium depletion after surgery and alongside radiotherapy

Following a successful surgery with a good prognosis, radiotherapy may be supplemented with deuterium depletion according to the *C/C/C/Radther Protocol*.

Experience to date shows that patients who use DDW according to the *C/C/Radther Protocol* and regularly repeat the *C/R/1 Protocol* recommended for patients in remission, may significantly reduce the risk of recurrence and remain cancer-free in the long term. At this point, a reference is made to the studies on breast cancer patients [65]. Out of forty-eight patients in remission, only one died during the 221-year cumulative follow-up period.

The use of deuterium depletion during pre-operative radiotherapy and following a surgery

Patients may start consuming 105 ppm DDW simultaneously (or preferably before) with radiotherapy according to the *C/C/Radther Protocol*. During radiotherapy and following the last treatment, a 105 ppm concentration should be maintained for two to five weeks (depending on the duration of radiotherapy). If a tumor is not operable, deuterium depletion may be used for three to four more weeks, by consuming 85 ppm DDW.

The effectiveness of radiotherapy is evaluated weeks after, as the size of the tumor may reduce even weeks after the therapy. The optimal time of surgery may be reconsidered, as increasing the dosage of DDW (105 ppm to 85 ppm) following radiotherapy may result in further regression. Determining the optimal time of surgery may be especially important in cases when the only option to remove a tumor is a major surgery that adversely affects the patient's quality of life.

Rectal cancer

A patient was diagnosed with rectal cancer near the anal sphincter muscle. In this case, it was a question of whether the surgery can be performed without proctectomy. The patient started to consume DDW at the same time as the radiotherapy before surgery. No tumor was detected at the time of the surgery. Nevertheless, the surgery was performed and the affected area was removed, yet an ostomy was not performed.

Breast cancer

Thanks to the development of surgical procedures, removing breast tumors is most commonly done with keeping the breast. Nevertheless, there are some borderline cases when either the breast size or location makes it difficult to keep the breast. Combining deuterium depletion with conventional treatments (radiotherapy, chemotherapy, and targeted therapies) may improve the chances of a surgery that keeps the breast.

Brain tumor (not glioblastoma)

For brain tumors, every millimeter of brain tissue retained intact has special importance. If the tumor is not a glioblastoma in histological terms, then administering 105 ppm DDW is recommended combined with radiotherapy. Two to five weeks after the end of radiotherapy it is recommended to switch to 85 ppm DDW. If regression occurs (and if there is no urgent need for surgery), a further reduction of deuterium levels (in two phases: 65 and 45 ppm DDW) is recommended until the surgery.

Using deuterium depletion after a surgery alongside conventional treatments

If the information obtained during surgery or pathology tests indicates the necessity of chemotherapy, it is combined with radiotherapy as part of the aftercare. The supportive use of DDW is recommended but exploiting the full potential of an increased DDW dosage should be reserved for later. In addition to radiotherapy and during the entire period of chemotherapy, the use of 105 ppm DDW is recommended. Switching to 85 ppm DDW may be considered before the last treatment.

If a glioblastoma was surgically removed and the treatment of the patient continues while observing the internationally accepted Stupp protocol, then it may be advisable to deviate from the above recommendations. Given the poor prognosis for this aggressive disease and a potentially short period without progression, the consumption of 85 ppm DDW is recommended even during radiotherapy. If Temozolomide is used in combination with radiotherapy, an 85 ppm level should be maintained for two to three weeks after the last radiotherapy. After this, it is advisable to switch to DDW with 65 ppm deuterium, and after another two to three months to 45 ppm, and then to DDW-25. If Temozolomide is administered after radiotherapy, it is not recommended to wait until the end of treatment before reducing the dose of DDW, as is generally recommended, but to reduce the deuterium concentration every two to three months.

Using deuterium depletion combined with post-operative chemotherapy

Adjuvant chemotherapy following a successful surgery

Depending on the histology and size of the removed tumor, protocols usually prescribe chemotherapy as aftercare. As discussed earlier, the reason for this is that by the time the disease is detected, tumor cells breaking off of the primary tumor have already made it to the bloodstream and the lymph, migrating to distant parts of the body. An adjuvant, aftercare chemotherapy aims to destroy the tumor cells that made it to distant parts of the body, thereby preventing the formation of future metastases.

If the patient has a good prognosis, because the size of the surgically removed tumor was small (one to two centimeters), and it has not affected the surrounding lymph nodes, then that specific tumor may respond well to chemotherapy. In such cases, it is then not necessary to use deuterium depletion simultaneously with chemotherapy. It is enough to only start using deuterium depletion at the time of the last two treatments or when chemotherapy is over.

The synergistic effect of deuterium depletion and chemotherapy is not fully understood, but DDW is confirmed to mitigate the toxic effects of cytostatic treatments. To optimize therapeutic benefits, in the first half of the aftercare, the treatment should be focused on chemotherapy, hoping that it would significantly reduce the number of circulating or attached tumor cells (potentially forming metastases in the future) in the bloodstream. Integrating deuterium depletion into the treatment regimen at the time of, or following, the last chemotherapy treatment may have the additional benefit of more effectively eliminating a reduced number of tumor cells. The additional benefit of using conventional and complementary treatments simultaneously is that deuterium depletion may extend patient aftercare by months, all by using an active substance that tumor cells have not been introduced to (see: *C/C/Chem Protocol*).

As a complementary treatment, deuterium depletion may be used in the treatment of colon and rectal cancers with good results. Veterinary experience suggests that dogs and cats treated for rectal cancer have an exceptionally high (over 70%) response to Vetera-DDW-25. Over 50% of them show complete recovery. In human medicine, patients undergo successful surgery in a significant percentage of cases. Chemotherapy is used as aftercare for the disease not to recur. However, thirty to forty percent of patients develop distant metastases, later on, most often affecting the liver. Of the 247 patients with colon and rectal cancer who chose deuterium depletion as a complementary option, twenty-six started to consume DDW while in remission. Only three of them died during the cumulative follow-up period of 128 years. The relapse-preventing (preventive in general) effects of DDW are demonstrated by the specific case of a patient who had to be operated on three times within three years. Following the regular use of deuterium depletion (repeating a DDW course thirteen times over ten years), the patient remained cancer-free for fifteen years.

Using deuterium depletion alongside chemotherapy preceding a surgery

Chemotherapy is used to achieve operability

The successful treatment of cancer patients depends on the operability of the tumor. Therefore, if radiotherapy is not an option, chemotherapy may help improve operability. In such a case, consuming 105 ppm or 85 ppm DDW may have an added benefit, as it contributes to a faster and greater regression, and to achieve operability earlier than otherwise. It is not necessary to alter the dosage after two to three months if the patient is in regression. However, if the treatment itself is not sufficient to achieve operability, decreasing the deuterium levels of DDW may be rescheduled earlier. If a patient has successfully undergone surgery, it is recommended to increase the dosage and continue using deuterium depletion.

When used in this combination, DDW may contribute to a faster and more significant regression of the tumor, improve operability, and support chemotherapy, mitigating its toxic effects.

Using deuterium depletion alongside chemotherapy and (subsequently) radiotherapy

The tumor is not operable, due to its location, staging, or classification

Of the common, generally operable cancers (breast, prostate, colon, and rectal cancer), up to 10% of lung cancer patients are operable at the time of the diagnosis. As previously described, nine lung cancer patients in remission used deuterium depletion. During the 53.7-year cumulative follow-up period, no death was recorded. In 90% of lung cancer patients, as surgery is ruled out, chemotherapy and radiotherapy are the only viable options. In such cases, observing the *1/C/C/Chem Protocol* is recommended. Apply 105 ppm DDW until chemotherapy is over if check-ups confirm a regression after the second or the third treatment.

When using deuterium depletion as a complementary treatment, regular check-ups show better-than-expected results, but this is attributed solely to chemotherapy. (The median survival time of 304 lung cancer patients who consumed DDW was six times longer, forty-eight months, than that of patients receiving conventional treatments only.) Encouraged by the outstanding results, patients may receive further treatments, despite the severe side effects. In the future, the planned use of DDW could contribute to the optimization of chemotherapy. Current scientific results and clinical data on deuterium depletion suggest that the cytostatic treatment could be used with the same or higher efficacy, coupled with lower toxicity. This consideration is not something that can be easily overlooked when it comes to the quality of life of patients.

To demonstrate how deuterium depletion may be combined with chemotherapy and radiotherapy, we present the example of a patient with a head/neck tumor. A 31-year-old female patient was diagnosed in October 2009 with laryngeal cancer and laryngectomy was recommended. The patient did not consent to the surgery. After the diagnosis, she started to use deuterium depletion and took advantage of the chemotherapy and radiotherapy offered. Complete regression of cancer was detected three months later, at the end of the treatments. The first five-month DDW course was followed by a one-month interruption. She then consumed DDW for another four months. These four months were followed by another three months of interruption. The four-month course was repeated twice with half-year interruptions. According to the last updates on her condition in January 2019, she had an intact voice box and larynx and had no complaints.

Using deuterium depletion alongside conventional treatments in recently diagnosed stage III patients

Recently diagnosed patients are in a more favorable position compared to those who transitioned into stage III despite years of conventional treatment and only considered deuterium depletion after. In this situation, deuterium depletion aims to serve as a complementary therapy to conventional treatments. The potential benefits of deuterium depletion should be reserved for a time when conventional treatments have already been completed and are no longer expected to bring more therapeutic results.

Deuterium depletion is recommended to start with 105 ppm. This deuterium level should be maintained throughout chemotherapy until the results from check-ups show a regression of cancer.

It has a measurable effect when tumor cells are first introduced to active substances and treatments. Deuterium depletion is capable of increasing the effectiveness of conventional treatments while mitigating its side effects. The permitted dosage of radiotherapy and the number of radiotherapy treatments are limited due to their severe side effects. In most cases, they are not sufficient to achieve a full recovery.

If patients have received the conventional treatments as prescribed by the protocols and there is a demonstrable improvement, deuterium depletion with a concentration of 85 ppm may continue for two to three months. After that, patients may switch to 65, 45, and finally, 25 ppm, respectively, to achieve a completely cancer-free condition. If intermittent follow-up examinations do not confirm the effectiveness of conventional treatments, the deuterium concentration should be reduced to 85 ppm during the period of the conventional treatments. In this case, the synergistic effects of combined therapies may result in the improvement of the patient's condition.

The use of deuterium depletion combined with oncological treatments in stage III patients receiving conventional treatments

Life expectancy is the worst for this group of patients, some of whom have never achieved a cancer-free condition, with others having achieved a cancer-free condition in the past but later developed distant metastases. The progression of cancer cannot be halted in the majority of patients, even despite the treatments with adverse side effects. The option of deuterium depletion was only considered in the late stage.

At this stage of cancer, deuterium depletion is only a last resort. Deuterium depletion should start from 105 ppm; however, after a few weeks of treatment, it may be medically indicated to switch to 85 ppm DDW. The duration of further steps should be shortened to one to two months. If the patient is in a critical situation, treatments may start with 85 ppm or even 65 ppm.

Several aspects should be considered in these cases.

If the dosage of DDW is inadequate, we may lose the chance to influence the course of the disease. If, however, the dosage is too high, then excessively rapid tumor necrosis and the related complications may cause problems in stage III patients. A further dilemma is whether to expect improvement from ongoing oncological treatments, in which case deuterium depletion may be reserved for a period when no further conventional treatments are available. This is why we recommend the use of a DDW of 105 ppm in the first few weeks alongside ongoing conventional treatments, as it will determine how sensitive the cancer is to deuterium depletion. If the treatment is effective, the 105 ppm level can be maintained, but if there is no significant improvement in the patient's condition and checkups show no evidence of regression, it is recommended to switch to a lower deuterium level (85 ppm). This should be followed by a further reduction in deuterium concentration every few months.

Use of deuterium depletion alongside or following targeted therapies

With the advances of molecular biology, genetic research has helped identify hundreds of genes whose mutations are associated with the development of cancer. Many signaling pathways that are damaged in cancer cells have been mapped. This has led to the development of targeted and conscious drug development, searching for and testing molecules that interact with the proteins encoded by faulty genes to inhibit the proliferation of cancer cells. Great hopes were pinned on drugs that inhibited tyrosine kinase because several cancers had been shown to have this enzyme malfunctioning in their cells. There are now several tyrosine kinase inhibitor drugs on the market, but these have only achieved significant results in treating/curing chronic myeloid leukemia (CML). The developers of tyrosine kinase inhibitors traced back the effect of these drugs to the inhibition of said gene. Researchers approaching cancer from the cellular metabolic perspective traced back the anti-cancer effect of the tyrosine kinase inhibitor named Gleevec to the stimulation of mitochondrial function. Stimulating the mitochondrial function also facilitates the production of low-deuterium metabolic water in the cells, thereby ensuring the healthy functioning of cellular processes [72]. There is ongoing drug development based on the signaling pathways identified in tumor cells. Despite encouraging results, there are limitations and risks of targeted therapeutic approaches, and the much-expected breakthrough has not yet been achieved.

While there are doubts about the applicability and sustainability of targeted therapies, some targeted therapeutic agents have lived up to expectations. There have been no targeted clinical trials to determine the extent to which the efficacy of these agents can be enhanced by deuterium depletion, but these trials should be reproduced in the future. The population followed included patients who received a targeted treatment alongside deuterium depletion. An example is given to illustrate the experience with the combined use of therapies.

Herceptin/Breast cancer

A 52-year-old HER-2 positive female patient was diagnosed in March 2004 with an inoperable inflammatory invasive breast tumor metastasized to the diaphragm. The patient then consumed DDW 105 ppm starting in July 2004, in addition to conventional treatment (chemotherapy, radiotherapy). Cancer responded well to treatment, detaching from the muscle, and became operable by October 2004. The patient continued chemotherapy treatment until February 2005, followed by Herceptin treatment for two years. The patient remained symptom-free until 2013 when she developed metastases in her bones and adrenal glands. The adrenal metastasis was removed along with the kidney, and the patient was again treated with chemotherapy and then hormone therapy, with bisphosphonate for the bone metastasis. Between 2004 and 2020, the patient consumed DDW twelve times, with continuous courses of nine to eleven months at the onset of the disease, followed by annual courses of two to three months. The importance of regular DDW consumption for maintaining a negative status was also demonstrated in her case, as after increasing the duration of the usual nine-month break to seventeen months, tumor marker levels started to rise. The trend was reversed by January 2020 after the resumption of the DDW regimen, and the patient was still in good overall health at the time of writing, with no complaints sixteen years after the initial diagnosis.

Gefitinib/Lung cancer

A 70-year-old male patient was diagnosed with lung cancer in September 2016, following a pathological fracture of his humerus. The patient started consuming DDW at 105 ppm two weeks after the diagnosis, with radiotherapy and medication for the bones starting in late October. Genetic testing identified an EGFR21 mutation in the patient and he started taking the drug Iressa in November. In October, the patient coughed up thick phlegm, and by November his bone pain had resolved, and he was able to work again in May 2017. CT scans in May and October of 2018 showed a slight progression in the lungs, the patient had stopped taking DDW before both scans. After the October CT scan, Iressa was withdrawn, and the patient was scheduled to continue treatment with immunotherapy.

Sutent/Kidney cancer

A male patient was diagnosed with clear cell renal cancer carcinoma in autumn 2003, which was surgically removed. Metastases appeared in autumn 2004 and again in August 2005, which were operated on each time. Another metastasis appeared in spring 2006, which progressed despite the treatment with medication (Sutent). This was confirmed by CT scans taken in March and August 2007, which showed two large metastases measuring 41 × 45 × 46 mm in the liver and one measuring 40 × 34 × 30 mm attached to the abdominal wall. The patient started consuming DDW in September 2007, and significant disease regression was described in May 2008, which continued according to the results of follow-up examinations in October 2008. The patient had been taking the drug Sutent since October 2006, but his disease progressed until autumn 2007 when he started deuterium depletion. After this, the regression of the disease occurred. The patient continued to take DDW for sixteen months, followed by a brief interruption and three more months of consuming DDW. Following another interruption, the patient consumed DDW for five more months before discontinuing it completely. An MRI scan taken one year later showed no change in the disease. A drastic deterioration in the patient's condition occurred three years later, in August 2012. The patient died in October 2012, nine years after the disease was discovered.

The above cases illustrate that combining certain targeted drugs with deuterium depletion may significantly improve the patients' life expectancy and highlight that carefully designed trials can optimize the combination of the two.

The combined use of deuterium depletion and immunotherapy

It is without a doubt that the immune system plays a crucial role in preventing and combating cancer, but the pharmaceutical industry's developments do not reflect this enough. There is a certain contradiction: the immune system's role in combating cancer is well known, yet patients are still treated with anti-cancer drugs that have severe adverse effects on the immune system. In patients using deuterium depletion, it was frequently observed that deuterium depletion protected the patients' hematopoietic system to some degree. Cytostatic treatments did not have to be stopped (or suspended only in very rare cases). There was no need to wait for weeks between two treatments until the blood test results went back to normal. The results from Phase II clinical trials conducted with diabetes patients confirmed that the counts of formed elements of blood increased during the three months of DDW-105 consumption. [44]

In many cases, patients want to support the immune system by the intake of high doses of vitamins during treatments. In accordance with scientific literature, it is not recommended to consume vitamins (especially antioxidants) in large amounts, when applying deuterium depletion, as they may significantly deteriorate the expected outcome of treatments, even worsening the prognosis of the disease [73].

Consuming beta-glucan polysaccharides, on the other hand, may have beneficial effects as they enhance the immune system. We, therefore, promote the consumption of medicinal mushrooms, as the beneficial effects of these mushrooms are attributed to polysaccharides.

CHAPTER NINE

Factors Affecting Dosage and the Efficacy of Deuterium Depletion, Findings Concerning the Usage of Deuterium Depletion

1. Daily DDW intake

Experience over several years has shown a clear correlation between the daily amount of DDW and the degree of impact. The higher the daily amount of DDW consumed and the lower the deuterium content, the greater the reduction in deuterium concentration in the body, with a proportional physiological and therapeutic effect. It is important to determine the dosage based on the actual condition. Patients should consume DDW with sufficient and necessary deuterium levels to achieve the desired results while maintaining a constant reduction of deuterium levels as long as possible.

2. Deuterium concentration of DDW

The lower the deuterium concentration of DDW, the more significant the reduction in deuterium levels in the body and the more pronounced the effect of consuming a unit volume of DDW. However, it does not follow that the lowest possible deuterium content of water should be offered to the patient right from the beginning.

3. The body's response to deuterium depletion

Based on the research results, veterinary and human experience of the past decades, there is reason enough to say that deuterium depletion has no toxic effects and its consumption does not have harmful unexpected side effects. If healthy people or asymptomatic patients in complete remission consume DDW, there are no noticeable signs. Using deuterium depletion may have subjective sensations as well as objective phenomena. Both are results of the interactions between the tumor and DDW. These sensations and phenomena provide useful pieces of information, which is why it is important to keep track of how patients feel. One of the most common symptoms is drowsiness and fatigue which appear one to two weeks (or sometimes later) after the start of DDW consumption. This symptom usually subsides within a few weeks, and then completely disappears. The length of this period depends on several factors. When establishing the dosage, it is of utmost importance to reduce the daily intake by ten to twenty percent if the patient's response to DDW is quick and strong. However, reducing the dosage is recommended only if using DDW causes strong discomfort for the patient. It is not recommended to completely suspend the use of DDW upon the disappearance or reduction of symptoms. If the symptoms subside, the dosage may be increased to the initial values.

There may be a reverse situation, in which deuterium depletion does not trigger any perceptible changes. In this case, if using deuterium depletion does not have any noticeable effects within one to two months, the dosage should be increased. This means raising the daily intake of DDW or using DDW with 20 ppm lower deuterium levels.

It is important to note that drowsiness and fatigue are the most likely to occur in the late stages of cancer, and patients with tumors three to six centimeters in size. If a tumor is smaller in size, the above symptoms may be milder or may not even be present despite the efficacy of DDW.

Veterinary experience has also confirmed the above; animals respond well to Vetera-DDW-25, tolerate the treatment, and the prostration and drowsiness that occur during the administration of the therapy usually diminish or disappear within two to three weeks.

4. Body weight

The most ideal subjects for DDW are small animals for veterinary use and children for human use, where high dosage of DDW can be administered due to their relatively low body weight. Taking the application of 105 ppm DDW as a basis, the daily fluid intake is in line with the optimal dosage of DDW (DdU - 1) for up to sixty to seventy kilograms of body weight. For a patient with a body weight of eighty to one hundred kilograms, increasing the daily fluid intake to 2 liters or more, 105 ppm may be suitable when starting the treatment, but a lower deuterium concentration will be required to increase the dosage.

5. The type and histological classification of the tumor

Based on our research and clinical experience, tumors of different origins and histology show different sensitivity to deuterium depletion. Due to the nature of results from mainly follow-up studies (in addition to Phase II clinical trials), it is hard to make a precise distinction in the case of the approximately 70 types of cancer. On the other hand, it is possible to reliably identify those patient groups that respond to treatment and those that do not. The majority of cancers fall into the well-responding group, with the following tumor types being particularly sensitive to deuterium depletion: breast, lung, stomach, prostate, kidney, bladder, ovarian, cervical, endometrial, testicle, tongue, larynx, and thyroid cancer, as well as ALL, AML, CLL and CML among leukemia. Also responding to treatment, but somewhat less sensitive compared to the above cancers are colon and rectal cancers, astrocytoma-type brain tumors, Hodgkin and non-Hodgkin lymphoma, and multiple myeloma. The cancers that are least sensitive to deuterium depletion (and conventional therapies), have the poorest prognosis, and are the most difficult to treat are glioblastoma multiforme (GBM), melanoma malignum, pancreatic and gallbladder cancers.

Thanks to the international advances in deuterium depletion research, a growing number of clinical tests, and the ever-growing corpus of results in the field, a significant enhancement in the effectiveness of deuterium depletion can be achieved, even for aggressive cancers. Back in 1999, when my book *Defeating Cancer!* was published, pancreatic cancer was mentioned as a type of tumor resistance to deuterium depletion. In 2014, at the annual conference of the American Association for Cancer Research (AACR) in San Diego [67], we could present a median survival of 39 months in the group of patients who were treated using deuterium depletion within sixty days of the diagnosis. In the control group receiving conventional treatment only, median survival was 6.8 months in female patients and only 5.8 months in male patients.

Some other cancers are mentioned for which it is still too early to take a stance on the effectiveness and sensitivity of deuterium depletion. These cancers have also mostly responded to deuterium depletion, but their classification is not yet established due to the low number of cases: tumors in the hard palate, pharyngeal, esophageal, and bone cancers, various types of soft tissue sarcomas, and neuroblastoma.

6. The tumor mass

There is also an evident correlation, as the smaller the tumor mass, the more effective the deuterium depletion treatment. In terms of dosage, this means that for a small tumor mass, a lower dose is sufficient to achieve regression. Also, a small tumor mass (as its necrosis does not have a substantial effect on the body) allows for a higher dose of DDW right at the beginning of the treatment. When determining the dosage in case of an advanced stage, the longest possible duration should be considered for the reduction of deuterium levels, alongside the principle of providing effective treatment.

7. The shape of the tumor and the impact on the surrounding tissues

Experience so far indicates that the part of the tumor in contact with healthy tissue is the most sensitive to deuterium depletion. This is due to the fact that it is the most invasive part of the tumor, with the highest number of dividing cells, which are in the most sensitive phase of the cell cycle to deuterium depletion. (Studies on cell cultures have shown that low deuterium concentrations prevent cells from moving from the G1 phase of cell division to the S phase.) Experience shows that the consumption of DDW first causes the infiltrating protrusions of the tumor surrounding tissue to recede, as evidenced by more than two decades of experience with the veterinary medicine Vetera-DDW-25.

This is of special significance in cases where the preoperative treatment of patients is possible. In these cases, the consumption of DDW may enhance operability, sparing the patients from larger-than-necessary surgical injuries, surgical and other complications. For brain tumors, the preventive consumption of DDW for three to four weeks may be particularly important, unless contraindicated for other reasons.

Also, tumors with a large surface area (for example, pleural tumors) are more sensitive to deuterium depletion than compact, large, and cohesive tumors.

8. The location of the tumor

A DDW treatment is the most effective when it comes into direct contact with the tumor and causes a substantial reduction of deuterium levels locally. Examples of such cancers are gastric, oral, or skin tumors or tumors and metastases located close to the skin surface. For this reason, for a patient with gastric cancer, fifty to sixty percent of the dosage calculated based on body weight may be sufficient at the start of the deuterium depletion treatment. For cancers of the oral cavity, it is recommended to keep DDW in the mouth for five to ten minutes before ingesting it and to repeat this process six to eight times a day. For tumors close to the skin surface, use gauze or cotton compresses impregnated with DDW. Local, external treatments are recommended to take place alongside the consumption of DDW.

9. Sensitivity of the tumor to deuterium depletion

When establishing the dose, it should be taken into account that while for sensitive tumor types, deuterium depletion is recommended at 105 ppm. For less sensitive, aggressive cancers (pancreatic cancer, glioblastoma, melanoma, sarcoma), it is recommended to start treatment with a lower, 85 ppm deuterium concentration. For cancers resistant (or less sensitive) to deuterium depletion, use 65–85 ppm DDW.

10. Treatment of the primary tumor and/or metastasis

It is generally true (albeit with significant differences) that primary tumors and metastases are sensitive to deuterium depletion. When establishing the dosage, the total tumor mass is the primary decisive factor, which may be more significant if metastases are already present. In some breast tumor patients with liver metastases, it was found that complete regression of the metastasis occurred sooner than that of the primary tumor. Lung metastases may also be sensitive to deuterium depletion but may require a longer time (eight to ten months) to take effect, while complete regression within one to two months has been recorded in liver metastases.

We also mention the case of two patients with gallbladder cancer where the primary gallbladder tumors were found to be sensitive to deuterium depletion, while their liver metastases were resistant to it.

11. Classification of the tumor stage at the beginning of deuterium depletion

The earlier deuterium depletion is integrated into the treatment regimen, the more successful the treatment can be and the better the chances of success. Consequently, if patients are subjected to deuterium depletion immediately after surgery and/or conventional treatment (when the number of circulating tumor cells is supposedly the lowest in their system), then it offers an opportunity to significantly reduce the chance of new tumors (recurrence/metastases) occurring.

12. General physical condition of the patient

If the patient's physical condition is satisfactory, the dosage may be established following the guidelines. In a patient with a deteriorated physical condition, the use of deuterium depletion may initially cause subjective deterioration, and the patient may become more debilitated and fatigued. This does not necessarily indicate the progression of the disease. The symptoms are caused by tumor necrosis. This period may last from one to two weeks to several months.

13. Other treatments

The vast majority of patients who used deuterium depletion as a complementary treatment also received conventional treatments. The exceptions were those for whom conventional treatment options were exhausted or not an option. In general, deuterium depletion has a synergistic effect when combined with conventional treatments, but the effect of deuterium depletion may often be impaired by the side effects of these treatments. The side effects of chemotherapy often lead to patients being unable to take in much liquid for several days, suffering from severe nausea and vomiting, and often stopping the consumption of DDW. However, in many cases, we found that patients who take DDW tolerate cytostatic treatments better, mainly because their blood counts do not deteriorate as much as expected and other side effects are less prevalent. In this case, cytostatic and radiotherapy treatments do not interfere with establishing the dosage. After surgery, DDW consumption may be suspended for four to six days, as during this period a larger amount of normal deuterium-containing liquid is introduced into the patient's body (in the form of an infusion), which cannot be compensated for by oral DDW intake.

Before officially recognizing the efficacy of a drug, in this phase of drug registration, deuterium depletion may only be a complementary treatment to human oncotherapies.

14. Complete blood count

No direct studies have been conducted to explore the relationship between immunological status and deuterium depletion, but experience showed that recovery rates with normal white blood cell counts were better than that of those patients whose blood counts had been negatively impacted by cytostatic treatment. In one patient with lung cancer, only stagnation was achieved, even after months of conventional treatment. Significant regression, despite continued consumption of DDW, occurred only months after cytostatic treatment had ended and the patient's blood count had normalized.

Clinical observations have shown that the blood counts of patients undergoing chemotherapy with deuterium depletion were better than expected, as confirmed by the results of the Phase II clinical trial in diabetic patients. In the study, thirty patients consumed DDW at 104 ppm for ninety days and blood counts (red blood cells, white blood cells, and platelets) improved significantly, but still within the normal range [44]. (Table 17)

Parameter tested	Day 1	Day 90	p value
White blood cells	4.52 10^{12}/L	6.96 10^9/L	0.01
Red blood cells	4.52 10^{12}/L	4.66 10^{12}/L	0.0064
Hemoglobin	8.36 mmol/L	8.61 mmol/L	0.011
Platelets	251.10^9/L	269.10^9/L	0.007

TABLE 17
Blood counts from thirty patients who consumed 104 ppm DDW for ninety days

15. Time elapsed since the start of deuterium depletion

As discussed before, one of the main considerations for establishing the correct dosage is that it is recommended to use lower concentrations of deuterium as the treatment progresses. As more time elapses since the start of the DDW treatment, the concentration of deuterium in the body decreases. This also means that some of the cells in the tumor may have adapted to the lower-than-natural deuterium environment. The length of the adaptation period depends on the type of cancer and may range from a few months to half a year or even a year. If no further reduction in deuterium levels can be induced, the aim is to keep deuterium levels low, which alone may be sufficient to ensure further improvement. However, cancers that adapt rapidly to low deuterium levels may be expected to progress again when there is no further reduction in deuterium levels in the body.

CHAPTER TEN

General Advice on the Application of Deuterium Depletion

1. Additional procedures that counteract deuterium depletion

Several patients also use some of the complementary therapies alongside conventional treatments, intending to improve their chances of recovery. Some of these methods have been shown to impair not only the effect of conventional therapies but also deuterium depletion.

The following additional methods and preparations are worth mentioning:

- Gerson diet
- Vegan diet
- Juice fasting
- Coenzyme Q10
- High doses of antioxidants (vitamins A, C, E, selenium)
- Hot tubs, sauna
- Iron supplements
- Intense, prolonged physical exertion, sports activities

We do not wish to comment on other methods and procedures used by patients. In general, it is considered that methods that inhibit the cell cycle at a certain point may make the tumor cells resistant to deuterium depletion.

2. How to consume DDW?

The daily amount is recommended to be consumed in single servings of 200–250 milliliters (6.8-8.5 fl.oz.). It is advisable to start and end the day by consuming DDW, and to consume additional amounts evenly throughout the day. DDW should be consumed ten to fifteen minutes before meals, and the normal deuterium concentration of food may be compensated for by DDW consumed after meals.

It is important that patients cover the majority of their daily fluid intake needs, at least 75–80% of it, with DDW.

This allows, for example, having a soup (made with normal water) and the consumption of vegetables and fruits. It is not recommended to consume larger amounts of soft drinks, fruit juices, milk, and other liquids that have a normal deuterium concentration. (Tea and coffee can be prepared using deuterium-depleted water.)

In the case of prematurely stopping deuterium depletion, the increase in deuterium concentration ceases the inhibition of tumor cell division, which may cause tumors to grow again, which, in turn, may be detected by diagnostic tools in the following weeks or months. It is to be assumed that tumor cells may benefit from a temporary increase in deuterium levels, for example, even for a few hours after a meal, due to the normal deuterium content of the food consumed. (This is particularly true for carbohydrates, which contain the highest amounts of deuterium.) Therefore, in more severe cases (a significant number or large size of metastases, aggressive cancer), it is recommended to consume DDW with a deuterium concentration 20–40 ppm lower than the currently used one. This helps compensate for the higher deuterium concentration of food. This prevents the creation of a favorable environment for the tumor up to several hours, significantly enhancing the efficacy of deuterium depletion.

3. How does the deuterium concentration of DDW vary when boiled and kept in the open air?

The composition of DDW does not change significantly during a single, short boil, so it may be used for brewing tea, coffee, or even cooking. The deuterium content of DDW increases measurably when it is mixed with normal water or stored in an open-air container for long periods. The rate of this process is illustrated by an experiment in which a few milliliters of DDW were left in the open air in a container. The deuterium content of the DDW was only 3 ppm higher 100 hours later, which implies that the deuterium concentration of DDW poured into the glass and left in the open air does not change significantly within a few hours. Nevertheless, it is recommended that DDW should be boiled within the shortest time possible in a closed container (under a lid) so that it does not come into direct contact with air (diffusion is faster at higher temperatures).

4. On the carbonic acid content of waters

Preventa deuterium-depleted drinking water is available in both carbonated and non-carbonated versions. Some people prefer the carbonated and others the non-carbonated versions, but there may be circumstances that contraindicate the consumption of carbonated drinks. These include the presence of oral cavity or stomach tumors. If only carbonated versions are available, in such cases it is recommended that the carbonic acid is removed by boiling, taking into account the considerations described above.

It is often said that carbonated waters are harmful to health because they cause the "acidification" of the body. It is important to know that carbonated and non-carbonated water contains roughly the same amount of minerals, so there are no significant differences between the two types of water in terms of acid-base balance. The stability of the blood pH in the body is ensured by a complex buffer system, which means that the 900-1000 grams of carbon dioxide produced every day as a result of normal metabolic processes do not cause the blood pH to shift towards acidity. Thus, the four to five grams of carbon dioxide in carbonated water do not have a significant effect, especially if we consider that a significant part of the carbon dioxide disappears immediately when we open a bottle and pour the carbonated liquid into a glass. Air bubbles are also released from the stomach through the mouth. There have been no findings in the previous years that would suggest that the carbonic acid content of water affects the efficacy of deuterium depletion. Carbonated waters may even have a beneficial physiological effect in warm weather by stimulating blood flow and improving heat dissipation.

5. For how long should DDW be consumed?

An important consideration in determining the duration of the treatment is that deuterium depletion has no adverse effects that would limit its applicability. The issue of when to discontinue the treatment is more important, as cancer progression may occur even after two to three months of discontinuation, so it is not advisable to discontinue DDW too early. Even if a patient is considered cured, it's recommended to continue the use of DDW for at least two months, but preferably four to six months. An important task in the near future is to determine, through studies, and aftercare strategy with high chances of recovery for patients. Experience so far has shown that it is safe to repeat the regimens on an annual basis, as described in the *Advice on dosage* section.

6. Interrupting a deuterium depletion course

In some cases, the patient may have interrupted taking DDW for shorter or longer periods. Patients then showed signs of cancer progression within a short time. In general, the process is reversible. If the effect was demonstrable before the interruption, then deuterium depletion may be continued effectively. After repeatedly resuming a DDW treatment, effectiveness may be reduced. Hence, one of the key rules of establishing the dosage is to avoid interrupting the intake of DDW before a complete regression is achieved.

7. How to end a DDW course?

Experience gained over the recent years has made it possible to abruptly switch from DDW to normal-deuterium liquids. Earlier recommendations suggested a slow and gradual increase of deuterium levels back to normal. Ideally, a DDW course is over when a patient has achieved a cancer-free condition. However, in many cases, the consumption of DDW is discontinued while the tumor is still present in the patient's body. If patients suspend the consumption of DDW, a sudden switching to normal-deuterium water results in further improvement of the patient's condition in some cases. This suggests that not only deuterium depletion, but also an abrupt increase of deuterium concentration, may have substantial effects on tumor cells. (It does not follow that it would be beneficial to raise deuterium concentration above 145–150 ppm, as preclinical studies have shown that this stimulated the proliferation of tumor cells.)

8. Long-term positive effects of deuterium depletion

The long-term follow-up (3, 4, 5, 10, and 20 years, respectively) reveals that no acute or chronic side effects were present. Some patients, however, interrupted the consumption of DDW after years of using it, even though they still had a tumor. The cases of these patients clearly showed that if the body's deuterium levels were kept low enough for some time, it had beneficial effects for the patients even when deuterium concentration increased to a normal level in the meantime. This observation is in line with previous experience that DDW is effective even if no further reduction in deuterium concentration is achieved in the body, but is maintained at a low level for a long period. The reason for this is that low deuterium levels have a major influence on cells' entire metabolism. Substituting deuterium in drinking water, as well as in nutrients for hydrogen, plays the same role that mitochondria in healthy cells play when producing low-deuterium metabolic water.

The above phenomenon has not been detected when patients suspend the consumption of DDW after a short period (a few months).

9. Diets supporting deuterium depletion

In our first published journal article [12] in 1993, we suggested that cells' deuterium/hydrogen ratio may change, and this change may fulfill an important regulatory role. The changes in the deuterium/hydrogen ratio have been confirmed in various molecules before the publication of our article. In *Chapter Two*, it was discussed that in molecules of plants, the deuterium/hydrogen ratio may substantially be different from that of water in the environment. This observation is explained by the differences in plant metabolism (C3, C4, and CAM plants). As a consequence, in C_3 plants (for example, spinach, wheat, rice, barley) the deuterium concentration of sugar molecules is 10–15 ppm lower than in C_4 plants (for example, corn, sugar cane, millet) [38, 39].

Other significant differences were observed in the deuterium concentration of foods, as reported in Chapter Two, "*A paradigm shift in biology.*" (See *Appendix* for an extended list of deuterium concentrations in foods.)

These results suggest that the isotope effect in biochemical processes, as well as the deuterium concentration of molecules involved in chemical reactions, determine the likelihood of the "heavy isotope" of hydrogen (deuterium) occurring at a given point of the compound that is being synthesized. The distribution and position of deuterium in different molecules (and the change of these properties) is predetermined and not random. This is how deuterium fulfills a regulatory role.

When creating a diet plan that supports deuterium depletion, those nutrients with lower deuterium levels are preferred. The data in Table 2 clearly shows that the deuterium content of carbohydrates is high. Therefore, the daily energy intake should focus on fats and oils, as these have lower deuterium levels. This nutritional approach is best reflected in the ketogenic diet, the therapeutic benefits of which have been proven by many clinical trials. Just by significantly reducing the intake of carbohydrates, patients achieve lower deuterium levels. This may partly explain the anti-cancer effect of the ketogenic diet.

If patients observe a vegan or vegetarian diet, it may impair the effectiveness of deuterium depletion. Similarly, the Gerson diet and juice fasting are not recommended when used alongside deuterium depletion. A further factor that may determine the effectiveness of DDW is that when a vegetarian diet is observed, the body's access to proteins is more restricted. The composition and proportion of amino acids from plants may be substantially different from that of animal proteins. This may partly explain why vegetarian diets can also slow down the tumor growth rate, as tumor cells do not have access to the amino acids they need to divide rapidly.

In the evaluations, particular attention was paid to cases where deuterium depletion did not result in an improvement, or the improvement was lower than expected. A review of these cases revealed that commonality was high antioxidant intake. This is also confirmed by independent studies carried out by colleagues at the Karolinska Institute [61]. According to their published results, DDW induces oxidative stress in cells, generating reactive radicals that trigger apoptosis (programmed cell death). Antioxidants, when consumed in large doses, (fulfilling their primary role) are capable of neutralizing free radicals, thereby protecting tumor cells and reducing the effectiveness of deuterium depletion, as well as conventional oncotherapy. Similarly, coenzyme Q_{10} and iron supplements deteriorate the effectiveness of DDW. These substances are likely to support electron transport in the mitochondria and maintain a redox balance inside the cells, therefore no sufficient amount of free radicals are created to initiate the apoptosis.

The measurements showed no significant difference in deuterium levels among different animal meats. Beef contains 138 ppm deuterium, pork 138 ppm, and chicken 137 ppm. Non-industrial meats with a higher fat content should be preferred. Of the dairy products, butter, ripe cheese, kefir, and yogurt should be preferred, especially those originating from the milk of grazing animals.

In terms of deuterium content, it should be noted that consuming fruits grown in the temperate climate zone is recommended. Closer to the Equator means an increasing deuterium concentration of precipitation, which manifests in the deuterium content of tropical plants and fruits. This does not necessarily mean that we should not consume bananas or oranges, only we should limit the consumption of these fruits.

The deuterium levels of fruit concentrates may also exceed that of drinking water, although many people do not attribute much importance to this difference. However, for some people, the deuterium levels of the foods consumed also have special importance, as it is crucial to reduce the deuterium concentration in their bodies. Any food containing high levels of deuterium has a countereffect on deuterium depletion, impairing its effectiveness. That is why it is recommended to limit or avoid the intake of such foods.

10. Other additional procedures

Patients and family members may have a hard time when introduced to so many potential remedies. They have many approaches and complementary treatment options, and it is hard to determine which one has solid scientific foundations and genuine benefits. The guiding principle should be to gather extensive information and choose two or three of the available complementary options to follow regularly. Persistence is the key to using these complementary therapies. When it comes to choosing complementary treatments, patients can make two typical mistakes. One possible mistake is that patients insist on using a given product, despite it being ineffective. The other mistake is prematurely discontinuing treatment when the results are not even supposed to be observed. There is no absolute, decisive truth. Therefore, there should be no determined timeline for any treatment. It is important to exercise caution when choosing complementary treatments and using them reasonably.

Several patients used products based on wheat germs, medicinal mushrooms, or something else alongside deuterium depletion. It is impossible to say to what extent these agents contributed to the patient's improvement, but the observations suggest that these treatments may have enhanced rather than impaired the effects of one another.

The most important factor is using DDW alongside conventional treatments and not instead of them.

CHAPTER ELEVEN
The Most Common Accompanying Symptom of Deuterium Depletion

Deuterium depletion is used predominantly by those who have cancer. Only a few patients who achieved a cancer-free condition as a result of the treatments before consuming DDW continue to use it. In the latter case, DDW has no noticeable effects, except for a lower chance of recurrence (which is not something to be overlooked). Those with detectable tumors noticed some symptoms or changes. These symptoms and changes are due to the necrosis of the tumor cell mass and the related physiological processes (e.g., inflammation). Some phenomena have been observed more widely, while others have been observed in a single tumor only.

Weakness, prostration, drowsiness

These symptoms occur in most patients using deuterium depletion, usually as early as a few weeks after starting DDW consumption, and last for varying periods. There are no noticeable symptoms if the tumor is small (one to two centimeters in diameter), as necrosis does not place a significant burden on the body. This phenomenon has also been observed in dogs and cats with cancer. The animals usually just sleep and lie down for the first week or two of treatment. Weakness and drowsiness are attributed to the necrotizing tumor mass and the related physiological changes.

Increasing the dosage may also cause increased fatigue and drowsiness

If deuterium depletion is continued with low-deuterium DDW, then the symptoms (drowsiness, fatigue) experienced at the beginning of the treatment may reappear. This may be considered a positive sign, as a repeated decrease of deuterium concentration induces a response, resulting in the further regression of the tumor.

Blushing, increased temperature, and fever spikes

Fever and fever spikes were detected only in a late stage of cancer or in the case of large tumor masses. It is known that when a tumor mass reaches a certain size, it spontaneously undergoes necrosis (accompanied by fever spikes). This phenomenon was observed when using deuterium depletion and was traced back to the process of necrosis affecting a large tumor mass. Frequent blushing or local redness of the skin were often observed.

Intermittently increasing pain

Starting to use deuterium depletion does not mean that patients may feel better within a short period. The previously described symptoms are the results of the treatment, and it may be accompanied by an intermittent increase in pain. This is mostly true for bone metastases and is less likely to occur in tumors of the soft tissue. Consulting a doctor and receiving adequate help may reverse this temporary deterioration of the patient's condition and relieve the pain.

Pain management

It is difficult to predict whether the relief of the pain is preceded by a temporary increase in a particular patient. Nevertheless, these are sure signs of improvement in the patient's condition. The relief of pain is explained by an improvement in the patient's underlying condition. However, it should be noted that at the start of a deuterium depletion treatment, the severity of the pain may fluctuate. This temporary period may range from two to three months.

Swelling and softening of the tumor-affected area

In tumors close to the skin surface, it frequently happens that after deuterium depletion, the tumor temporarily grows, but becomes softer. Follow-up examinations using imaging tests detected an increased tumor size in several cases. All of these were detected a few weeks after starting the consumption of DDW. Therefore, it may be important to synchronize imaging tests with the consumption of DDW (as described previously). Continuing deuterium depletion may cause a significant regression and reduction in the size of the tumor. A temporary "growth" may be traced back to the inflammatory processes accompanying the application of DDW.

Local warmth of the affected area

In several cases, patients experienced the warmth of the skin surface affected by tumors. It is recommended to cool down the affected area, which reduces the inflammation or shortens its duration. Additionally, it is generally true that hot baths or saunas are not recommended during the consumption of DDW.

Cerebral edema

If the brain is affected, tumor necrosis poses an enormous challenge for the entire body. In several cases, it was observed that the symptoms of brain cancer patients became more severe as a result of DDW consumption. The phenomenon is explained by the temporary formation of edema due to tumor necrosis, or the size of extant edema grows, causing an increase of intracranial pressure. Patients therefore should be prepared for an "apparent" temporary deterioration in their condition. The edema should be treated with the available therapeutic tools.

Pulling and tingling sensation in the tumor

These symptoms are caused by the processes in the tumor and are the results of deuterium depletion.

Minor bleeding in the bladder, stomach, or rectum

The location of the tumor determines the symptoms of tumor necrosis. In the above three types of cancer, it occurred several times that necrotized tissues detached from the tumor and were released from the body. The process was sometimes accompanied by a minor hemorrhage.

An improvement of appetite and general health

At least fifty to sixty percent of patients experienced a better physical condition after an initial deterioration. Their appetite and general health also improved.

Weight gain

In addition to the above accompanying phenomena, weight gain was observed in several cases.

Exudation and wound healing in ulcerating tumors

Following the start of a deuterium depletion treatment, ulcerating tumors start to heavily exude. This should be interpreted as a positive sign, proven by the subsequent formation of a "crater" in the affected area. This "crater" later closes and the affected area heals. The phenomenon was also observed and published when using Vetera-DDW-25 deuterium-depleted veterinary anticancer medicinal product.

An improvement of general comfort

Patients feel better even a few weeks or months after starting a DDW treatment. Their general comfort and stamina improve.

Brick dust urine

In some cases, orange or red, brick dust-colored urine was observed. In some other cases, urine became cloudy and had a very unpleasant odor. If that is the case, patients should have their uric acid levels tested and consult a physician.

Better tolerance of radiotherapy and cytostatic treatments

Deuterium depletion synergizes with conventional treatments and (in addition to their anti-cancer effects) mitigates their side effects. The blood counts of patients using deuterium depletion did not deteriorate as would otherwise be expected. In some cases, severe nausea was not present. The consumption of DDW (and especially a high DDW-containing gel) mitigated skin irritation resulting from radiotherapy.

Transient coughing in lung cancer patients

In lung cancer patients, the effects of DDW may manifest in transient coughing. The intensity of coughing is related to the location and size of the tumor. Patients may cough up a white, viscose phlegm mostly in the morning. As the urge to cough subsides, the patients' respiration also improves. Occasionally, blood may be present in the phlegm, which is explained by damages to minor blood vessels.

If the phlegm is yellow or green, then it is recommended to use antibiotics based on a consultation with a physician, as it is likely that the disintegrating tumor acts as a growth medium for the bacteria.

Tumor necrosis may cause abscesses

A disintegrating, necrotizing tumor is an ideal growth medium for bacteria. It may cause severe complications for two types of cancer: lung and gastrointestinal cancers. In the case of the lung, necrotizing tumor tissue causes severe irritation. The resulting strong urge to cough helps remove necrotized tissues. In this case, if the phlegm is discolored (becomes yellowish-green), it is recommended to administer antibiotics to the patient. The passing of necrotized tissues from the gastrointestinal tract may also cause complications. In the case of the colon, rectal and gastric cancers, mucous secretions and small fragments of tissue are often passed in the feces. But in the case of pancreatic cancer, the necrotizing tissue may not enter the intestinal tract, which can lead to abscess formation. Similarly, abscesses may form if the size of the primary colon and rectal tumor is so large that its regression is delayed and the dead tissue fragments cannot detach and leave the body.

CHAPTER TWELVE

The Main Phases of the Application of Deuterium Depletion (1992–2020)

Access to DDW has substantially improved over the years, also affecting the effectiveness of deuterium depletion. The transition from one phase to the other was gradual. In some cases, the treatment of patients encompasses several phases. This chapter describes the effects of deuterium depletion through specific case studies. As discussed earlier, the application of DDW does not mean an immediate and complete recovery. In the majority of cases, patients started using deuterium depletion in a late stage only. Deuterium depletion was capable of slowing down, stopping, or even reversing the malignant processes (even if only temporarily). Over time, as the concentration range of DDW became broader, efficacy and the obtained results improved. Had we possessed the knowledge, experience and availability of low-deuterium DDW, we could have achieved better results even in the beginning.

The first phase (1992–1995)
The first patients who had access to DDW consumed it in concentrations of 90–100 ppm.

The second phase (1995–1996)
25 ppm DDW first became available in August 1995. In this phase, as opposed to the later dosage model, patients supplemented the consumption of 90–100 ppm DDW with a few hundred milliliters (8-10 fl. oz.) of 25 ppm DDW.

The third phase (1996–1998)
45–50 ppm DDW became available between October 1996 and April 1998. The first three phases were characterized by a limited amount of DDW, which was particularly true for the lower deuterium concentration range.

The fourth phase (1998–2000)
High-performance machinery was installed on the production line in May 1998. It made possible the production of 85 ppm DDW and a substantial amount of 25 ppm water. At this point, it became possible to schedule the changes in deuterium levels. The different dilution ratios of 25 ppm DDW enabled a gradual reduction of deuterium concentrations. This phase lasted until January 2000, with the commercial availability of Preventa–105 and Preventa–85 deuterium-depleted waters.

The fifth phase (2000 to today)
Following the commercial introduction through the distribution channels, Preventa–105 and Preventa–85, a relatively large number of patients used DDW, most of whom we do not have any information about. In some cases, patients contacted us and asked for advice after months of consuming Preventa. Overall, most patients consumed DDW in the range of 105–25 ppm between January 2000 and January 2020, with deuterium concentrations gradually decreasing.

CHAPTER THIRTEEN

A Demonstration of the Effectiveness of Deuterium Depletion through Case Studies

In previous chapters, the effectiveness of DDW was shown through the statistical analysis of large patient populations, comparing the data of patients using deuterium depletion to data from conventional treatments. The statistical analysis clearly shows the high efficacy of deuterium depletion, albeit provides no insight into the individual cases behind the figures. This is why we believe it is important to show, through concrete cases, the successes and failures that have led to the development of knowledge and information on the use of DDW. Based on the feedback we received on the book *Defeating Cancer!*, this section gave us the most hope and encouragement that we should not give up because there is a chance for a cure. I sincerely hope that many readers will recognize themselves in these examples, and it will give them strength and help in early recovery.

CASE STUDIES

Lung tumor

Age of the patient	Sex of the patient	Date of the diagnosis	Start of deuterium depletion	Last update
61 years old	Male	October 1992	April 1993	June 2006

A 61-year-old male patient was diagnosed with inoperable lung cancer in 1992 and subsequently underwent radiotherapy. He first consumed DDW regularly from April 1993 to September 1993, during which time no progression of the disease could be detected. Later, the patient discontinued the treatment due to the resulting pain after five to ten minutes of consuming DDW. Subsequently, the patient lost fifteen kilograms, until March 1994 when he resumed DDW consumption based on medical advice. In the following two months he gained four kilograms, and an X-ray in July showed that the tumor was the same size as two years earlier. His shortness of breath disappeared and he became physically more active. A check-up in January 1995 confirmed the stagnation of the tumor, and a bronchoscopy in June confirmed fibrosis in the affected area. Subsequently, the patient enjoyed good general health. An X-ray taken in May 1996 showed regression. The patient discontinued the consumption of DDW in August 1996 in good general health. His condition deteriorated again starting in January 1997. In May 1999, he experienced significant pain in the surgical area. However, he was physically active and maintained his weight. At this point, seven years had passed since the initial diagnosis and exploratory surgery. X-rays showed encapsulation of the tumor. In 2005, the patient was diagnosed with oral cavity cancer infiltrating the jaw. Surgery was performed and the patient received radiotherapy. At that time, he resumed the consumption of DDW, albeit in a small amount. The patient died in summer 2006, fourteen years after being diagnosed with inoperable lung cancer.

Age of the patient	Sex of the patient	Date of the diagnosis	Start of deuterium depletion	Last update
54 years old	Female	January 1994	February 1994	September 1998

A 54-year-old female patient was diagnosed with adenocarcinoma in January 1994. She started consuming DDW a month later and continued until October 1995 (twenty consecutive months). The disease did not progress during this period. By June 1995, the previously fist-sized tumor had become smaller. The patient enjoyed good general health when stopping the consumption of DDW. She continued working in September 1998, four and a half years after being diagnosed with adenocarcinoma. She had an active lifestyle. However, at that time, she developed a metastasis under the skin on the abdominal wall. No further information was received on the patient's condition.

Age of the patient	Sex of the patient	Date of the diagnosis	Start of deuterium depletion	Last update
61 years old	Female	September 1994	November 1994	August 1998

A 61-year-old female patient was diagnosed with small cell lung cancer in September 1994 and started cytostatic treatment in October. The patient responded to the therapy but had difficulty tolerating the side effects. She consumed DDW without interruption from November 1994 to October 1995 and was in good physical condition. The patient died three years after discontinuing the DDW course in August 1998.

Age of the patient	Sex of the patient	Date of the diagnosis	Start of deuterium depletion	Last update
75 years old	Female	June 1993	April 1995	September 2005

The patient underwent surgery for lung adenocarcinoma in August 1993. The tumor was attached to the pleura at the apex of the lung. In the center of the third lung segment was a walnut-sized tumor that was removed with lobectomy, along with hilum lymph nodes. Despite the successful surgery, the patient's CA 19-9 tumor marker levels started to rise, starting from late 1994. She started consuming 90 ppm DDW in April 1995. Subsequently, tumor marker levels increased slightly. Figure 20 shows the evolution of CA 19-9 tumor marker values between late 1995 and 1998. The patient continued to consume DDW in gradually decreasing D-concentration. The average daily deuterium concentration was 80 ppm in 1997, 70 ppm in 1998, 40 ppm in 1999, and 25 ppm in 2001. The patient led an active life and did not have any complaints until spring 2004 when she was diagnosed with gynecologic cancer with metastases in the liver and the lungs. She died in September 2005 as a result of cerebral infarction at the age of 85.

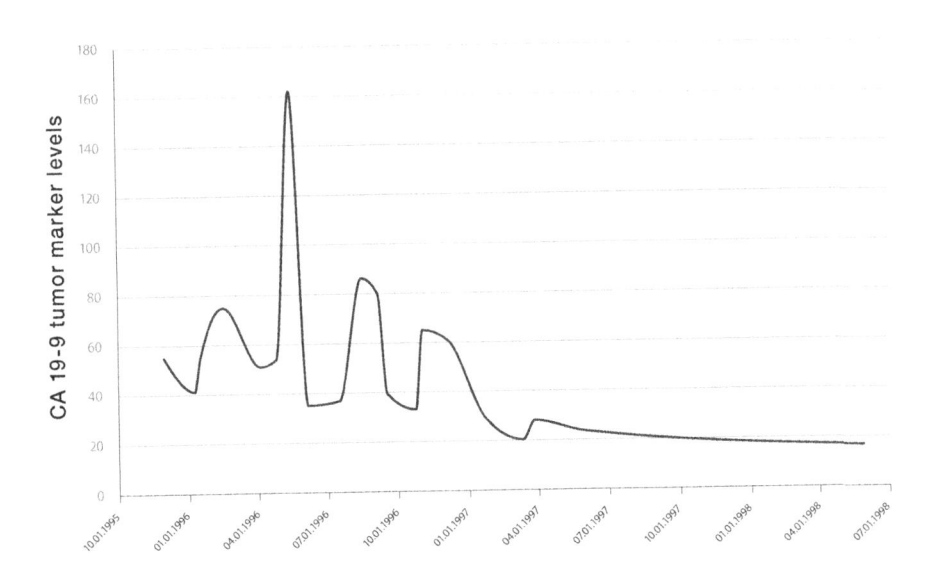

FIGURE 20
Evolution of CA 19-9 levels in a 75-year-old female lung cancer patient from the end of 1994 to March 1998

Age of the patient	Sex of the patient	Date of the diagnosis	Start of deuterium depletion	Last update
69 years old	Male	July 1995	October 1995	September 1999

Check-ups in June 1995 showed squamous cell carcinoma in the left lung. The tumor infiltrated into the pericardium and adhered to the great vessels. Conventional therapies were not an option due to the size and location of the tumor. Deuterium depletion was the only feasible tool to improve the patient's condition. After consuming DDW for half a year, he was in a good condition by April 1996. The patient's erythrocyte sedimentation rate (ESR) reduced from the previous eighty to six, with the size of the tumor not changing. The patient's blood sugar levels (he had diabetes as an underlying condition) and shortness of breath were reduced. Pleural effusion on the left lung was absorbed, and the patient could return to his work as a doctor. In autumn 1997, he could even travel abroad. His condition and general comfort were good until June 1998, when a slow growth of the tumor was observed. In July, the patient suffered a heart attack and had breathing difficulties even when resting. The patient's cardiac symptoms were kept under control with medication. The tumor showed minimal growth, even despite the discontinuation of DDW after the heart attack. The patient resumed the consumption of DDW in January 1999. In June 1999, the patient no longer experienced shortness of breath. X-rays showed a moderate progression of cancer. He consumed DDW with a single interruption until June 1999. The patient died three months later, at the age of seventy-four.

Age of the patient	Sex of the patient	Date of the diagnosis	Start of deuterium depletion	Last update
54 years old	Female	July 2001	July 2001	July 2012

A 54-year-old female patient was diagnosed with a brain tumor in July 2001. The tumor was later confirmed to be a metastasis of small cell lung cancer. The patient was administered chemotherapy and radiotherapy. However, in the doctor's opinion, she had a prognosis of only a few months. Following the diagnosis, she started consuming DDW within a week. A cranial MRI taken three months later confirmed regression, which continued throughout the next two years. An MRI in October 2004 showed no metastases (Fig. 15). Earlier, in February 2003, complete regression was observed in the lungs. The patient continued to consume DDW for the next two years with brief interruptions until 2005. Lung cancer recurred seven years later, in 2012, and chemotherapy was resumed. The patient consumed DDW for eight months, followed by two months of interruption, then resumed consuming DDW for another five months. The patient died in 2013, twelve years after the detection of the brain metastasis.

Age of the patient	Sex of the patient	Date of the diagnosis	Start of deuterium depletion	Last update
47 years old	Male	May 1993	January 1994	October 1996

A 47-year-old male patient was diagnosed in May 1993 with inoperable epithelial carcinoma of the lung. By the start of deuterium depletion treatment (January 1994), the patient had already lost fifteen kilograms. The size of the tumor stagnated with minimal regression, then in December 1993, progression was observed. In January 1994, the patient's treating physician switched protocol. The patient tolerated the treatments and his condition was described as stagnating. Occasionally, reduced atelectasis was reported (August 1994). The patient was administered the last treatment in September 1994. A routine check-up three months later, in December 1994, detected major regression. At this point, deuterium depletion was the only treatment option. The improvement continued as reported in March 1995. The patient consumed DDW until April 1996. He died half a year after this due to the progression of the disease, three and half years after the diagnosis.

Age of the patient	Sex of the patient	Date of the diagnosis	Start of deuterium depletion	Last update
58 years old	Female	November 2007	November 2007	January 2019

A 58-year-old female patient was diagnosed in November 2007 with a brain metastasis originating from a primary lung tumor. Life expectancy was estimated at only a few weeks. Alongside conventional treatments (chemotherapy, radiotherapy), the patient continued DDW starting from the treatment. A check-up a few months later reported the complete regression of the brain metastasis. The patient continued the intake of DDW for nearly three years, until August 2010. Then, after a four-month interruption, cancer progression was observed in the lungs. The patient then resumed the consumption of DDW, resulting in an improvement of her condition. Over the years, the patient observed eight deuterium depletion courses. On four occasions, the disease progressed and could be successfully halted. The patient was in a good general condition ten years after the diagnosis and died in January 2019 after a two-year interruption of DDW treatment.

Age of the patient	Sex of the patient	Date of the diagnosis	Start of deuterium depletion	Last update
70 years old	Male	September 2016	September 2016	February 2020

A 70-year-old male patient was diagnosed with lung cancer after a pathological fracture of the humerus in September 2016. Two weeks after diagnosis, the patient started consuming 105 ppm DDW, and deuterium depletion was complemented by radiotherapy of the bones and targeted therapy at the end of October. Genetic testing identified an EGFR21 mutation in the patient and he started taking the drug Iressa in November. In October, he coughed up phlegm from his lungs. By November, his bone pain had resolved and he returned to work in May 2017. In May and October 2018, a CT test showed minimal cancer progression in the lungs. The patient interrupted the consumption of DDW before both CT tests. Following the CT test showing progression in October, the administration of Iressa was discontinued. The treatment was planned to continue with immunotherapy. The patient consumed DDW until May 2019. The last reports on the patient's condition from February 2020 confirmed the appearance of minor metastases and a major regression of the lung tumor.

Age of the patient	Sex of the patient	Date of the diagnosis	Start of deuterium depletion	Last update
53 years old	Male	February 2010	August 2010	April 2018

A 53-year-old male patient underwent surgery for lung cancer in February 2010. During the surgery, a six-centimeter fragment was removed from the second rib (affected by the tumor). Following the surgery, the patient was administered Gemcitabine and Cisplatin. After the successful surgery, the patient did not start the consumption of DDW immediately. Deuterium depletion treatment was administered only after the fourth round of chemotherapy. In the six years after this, the patient observed seven DDW–25 courses, over a total of 600 days. All check-ups had negative results.

Age of the patient	Sex of the patient	Date of the diagnosis	Start of deuterium depletion	Last update
53 years old	Male	April 2006	December 2006	February 2020

A 53-year-old male patient underwent successful surgery in July 2006, then was administered chemotherapy. He first started to use DDW in December 2006, for six months. This initial course was followed by five more, over a total of eighteen months. The last check-up in 2020 showed negative results.

Breast cancer

Age of the patient	Sex of the patient	Date of the diagnosis	Start of deuterium depletion	Last update
46 years old	Female	October 1988	July 1993	May 1998

A 46-year-old female patient was diagnosed in 1988 with breast cancer, with a bone metastasis first confirmed earlier in September 1992. Examinations confirmed the progression of cancer until the start of the DDW course. The patient's pain intensified and increasingly had difficulty moving. Following the treatment with DDW, the patient's pain resolved within one to one and a half months. Bone scintigraphy two months later did not show any of the previously observed multiple, small metastases. The patient continued to be in good general condition and consumed DDW until January 1994, then interrupted it. Four months later, her condition significantly deteriorated and her pain intensified. Bone scintigraphy in September 1994 confirmed a moderate progression of cancer. The patient decided to continue with DDW, and bone scintigraphy in October 1995 showed no further metastases compared to the previous status. The disease progressed at a slow pace. In the first years of the consumption of DDW, the patient's quality of life did not change significantly. As a result of the bone metastasis, the patient suffered a pathological fracture of her humerus in December 1996, after which the femur had to be surgically reinforced. The humerus was fused and in the summer of 1997 the patient gained ten kilograms, and her pain was kept under control with medication. During the first four years of treatment, no metastases could be confirmed in any soft parts of the body. A CT scan in October 1997 showed a brain metastasis. The patient died five and a half years after the bone metastasis was observed, during which time she consumed DDW for four and a half years.

Age of the patient	Sex of the patient	Date of the diagnosis	Start of deuterium depletion	Last update
39 years old	Female	June 1986	September 1994	March 2001

A 39-year-old female patient underwent surgery for the first time for her walnut-sized breast tumor in 1986. The first recurrence occurred in 1987. Bone metastasis was first suspected in September 1993, however, check-ups only confirmed it in April 1994. A month before that, a lung tumor had also been confirmed. The disease progressed despite intensive conventional treatment. Morphine was used to manage her intense pain. Following the start of deuterium depletion, the patient was able to stop taking analgesic medication within a few weeks. In October, a tumor underneath her nose flattened and reduced in size. A chest X-ray in November confirmed the regression of the lung metastasis. In March 1995, bone scintigraphy also confirmed regression. In January 1996, complete regression was observed in the lungs and further regression was observed in the bones. Starting from November 1996, the progression of the patient's under-the-scalp tumors was reported. In January 1997, the patient's lungs were clear and she did not experience pain. She consumed DDW until spring 1998, then, after a half-a-year interruption, continued to use it from November 1998 to late 1999. The patient had no complaints in 2000 but reported having pain in early 2001. Despite resuming the consumption of DDW, she died in March 2001. The patient did not experience pain for five years after starting to consume DDW. She enjoyed a good quality of life when cancer was in regression.

Age of the patient	Sex of the patient	Date of the diagnosis	Start of deuterium depletion	Last update
48 years old	Female	July 1986	August 1995	February 2006

A 48-year-old female patient underwent sectoral resection of the left breast with axillary block dissection for invasive lobular carcinoma in the summer of 1986. Axillary lymph nodes were also affected. Following the surgery, she was also administered radiotherapy, but she rejected chemotherapy. In 1990, a local metastasis was removed from the left breast and the patient was given radiotherapy repeatedly. In August 1995, an extensive, inoperable recurrence was discovered that infiltrated the entire residual breast and was attached to the bone. At this time, combined cytostatic treatment was started according to a CMF regimen, with deuterium depletion also starting in August 1995. The patient had a regression of over 50%, continuing even after chemotherapy has ended. The patient continued the consumption of DDW from August 1995 to 2000 and had no complaints in this period. Subsequently, until 2005, she only occasionally consumed DDW, while also receiving conventional treatment. The patient died twenty years after the diagnosis in March 2006.

Age of the patient	Sex of the patient	Date of the diagnosis	Start of deuterium depletion	Last update
37 years old	Female	June 1993	February 1996	December 2004

A 37-year-old female patient underwent surgery for breast cancer and subsequently was administered radiotherapy. In 1995, bone scintigraphy confirmed a metastasis. Chemotherapy was started at this point and ended in October 1996. The patient consumed DDW irregularly (due to chemotherapy) between February and August 1996. She repeatedly interrupted the treatment but consistently adhered to it from August 1996. The patient gained three kilograms from August to October 1996, and no longer needed walking support. In a follow-up examination in 1998, the patient's condition stabilized, and significant calcification was reported in her bones. The flexibility of the hip joint improved and the patient could cope with strenuous physical activities (swimming, hiking). She consumed DDW almost exclusively between 1996 and 2000. However, she significantly reduced the daily intake from 2000 and did not consume it for several days in a row. In September 2000, brain metastasis was removed, and the patient rapidly recovered. An MRI scan taken seven months later was negative, similarly to the ones taken in 2002, 2003, and 2004. No information was received about the patient after this.

Age of the patient	Sex of the patient	Date of the diagnosis	Start of deuterium depletion	Last update
54 years old	Female	June 1983	March 1997	September 2007

A 54-year-old female patient underwent surgery in 1983, and the fact that lymph nodes were also affected was confirmed during surgery. In 1992, after an epileptic seizure, brain metastasis was discovered. The metastasis was subsequently operated on, and the patient received chemotherapy and radiotherapy. In August 1996, lung metastasis was confirmed. In March 1997, a CT scan confirmed metastases in the liver and the adrenal glands. The patient then started to use deuterium depletion in March 1997. In August 1997, five months after the start of the DDW course, the metastasis in the liver stagnated. Regression was reported in October and complete regression in June 1998. The ultrasound scan performed in October 1998 was negative in terms of tumors in the liver as well as the adrenal glands. In addition to the regression in the liver and adrenal glands, the lung tumor progressed by late 1997. The patient received radiotherapy and two cytostatic treatments. Following the treatments, the patient enjoyed a good physical condition and had no symptoms starting from April 1998. During the summer, a substantial amount of white, viscose matter passed through her airways when coughing. This is something characteristic of lung cancer patients using deuterium depletion. The patient consumed DDW for ten years. In the last two years, she has reduced her daily intake. The patient lived fifteen years while enjoying a good quality of life, despite the metastases in the brain, lungs, liver, bones, and adrenal glands.

Age of the patient	Sex of the patient	Date of the diagnosis	Start of deuterium depletion	Last update
44 years old	Female	June 2007	December 2007	February 2020

A 44-year-old female patient was diagnosed with breast cancer in June 2007 and had metastasis in a vertebra. She received chemotherapy and radiotherapy. The patient started consuming DDW half a year after the diagnosis. In January 2009, a PET/CT scan showed regression in the bones, however, subsequent examinations did confirm a tumor. The patient first consumed DDW for ten months without interruption, then had three-month courses of DDW with interruptions ranging from three to six months until 2012. The patient's general condition was stable, but a one-millimeter growth was detected after an eighteen-month interruption of the consumption of DDW. A biopsy confirmed that the lesion was a bone metastasis of the primary breast tumor. After a minimal progression of cancer, the patient repeated DDW courses as previously. At the time of writing, thirteen years after the diagnosis, she has enjoyed a good quality of life.

Prostate tumor

Age of the patient	Sex of the patient	Date of the diagnosis	Start of deuterium depletion	Last update
68 years old	Male	October 1992	October 1992	November 2003

A 68-year-old male patient was diagnosed with inoperable prostate cancer in October 1992. After starting the treatment with DDW, his difficulties urinating resolved within a few weeks. His PSA levels decreased before starting a Fugerel treatment. One month later, the tumor was no longer palpable and became operable. The patient (also a medical doctor), however, rejected the surgery. The patient drank DDW for a year and remained symptom-free for ten years. At this point, since the bones were also affected, the vertebrae collapsed, making the patient bedridden. He then resumed the consumption of DDW. He was able to leave his bed and lead an active lifestyle. This was followed by a deterioration of his condition and he died eleven years after the diagnosis, in November 2003.

Age of the patient	Sex of the patient	Date of the diagnosis	Start of deuterium depletion	Last update
66 years old	Male	September 1994	November 1994	March 1997

A 66-year-old male patient was diagnosed with prostate cancer that caused a complete ureteral obstruction. In September 1994, lymph node metastases were detected, and the patient's PSA levels were at 83.4 ng/mL. After the consumption of DDW, the patient's difficulties in urinating quickly resolved and then completely disappeared. Two weeks after the start of deuterium depletion, PSA levels in the patient's blood were 0.99 ng/mL. The measurement was considered to be erroneous, therefore it was repeated two weeks later, showing PSA levels of only 0.6 ng/mL. PSA levels dropped further down to 0.23 ng/mL by March 1995. The values measured in June 1996 and January 1997 were below the detection threshold. The patient first consumed DDW for nine months, then had two- to three-month courses in 1996 and 1997. No further information was received about the patient's condition.

Age of the patient	Sex of the patient	Date of the diagnosis	Start of deuterium depletion	Last update
60 years old	Male	March 2002	April 2003	February 2018

A 60-year-old male patient was diagnosed in March 2002 with prostate cancer. He had PSA levels exceeding 1,000 ng/mL and extensive bone metastasis. The patient received conventional treatment, and started DDW only a year later, in April 2003, when the PSA had already decreased to 15 ng/mL. The patient then repeated courses of five to six months for six years, with interruptions of a few months. After three years, bone isotope tests showed regression. The patient repeated three-month DDW courses between 2009 and 2015 with five to six-month interruptions. According to the last update on the patient's condition, he was in a good general condition sixteen years after the diagnosis. In the past years, he received conventional treatments supplemented with DDW courses. The changes of the patient's PSA is shown in Figure 21 below.

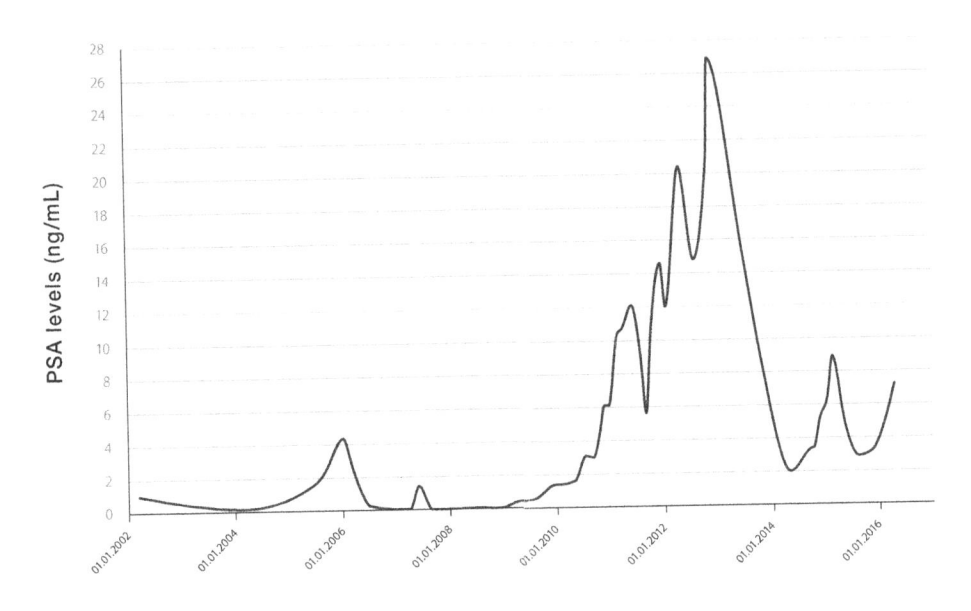

FIGURE 21
Changes in tumor marker levels in a patient diagnosed in 2002 with extensive bone metastasis and PSA levels exceeding 1,000 ng/mL, receiving conventional treatments

Age of the patient	Sex of the patient	Date of the diagnosis	Start of deuterium depletion	Last update
71 years old	Male	September 2005	October 2005	November 2011

A 71-year-old male patient was diagnosed with multiple bone metastases in September 2005 and had PSA levels of 540 ng/mL. The patient started to consume DDW one month after the diagnosis. Within one month, the PSA levels significantly reduced to 9.9 ng/mL. The patient repeated three months of DDW courses between 2005 and 2009, with three to four months of interruption. In April 2007, a bone isotope scan confirmed regression. An increase in PSA was detected from late 2007, which was associated with a pause in DDW consumption, as illustrated in the figure below (Fig. 22). The patient completed the DDW course in December 2007 with PSA levels at 4.6 ng/mL. By February 2008, PSA levels increased to 11.5 ng/mL. The increase in PSA levels continued during the first month after resuming deuterium depletion, up to 17.7 ng/mL. When the patient halted the DDW course four months later, PSA levels dropped to 9.9 ng/mL again. By the end of this interruption, PSA levels rose to 14.8 ng/mL, followed by a drop to approximately 10 by the end of a four-month interruption. PSA levels stabilized at 10. The patient discontinued the consumption of DDW in January 2009. Subsequently, PSA levels rapidly increased to 220 ng/mL in January 2011. Conventional treatments managed to reduce PSA levels to 40 ng/mL. According to the last update on the patient's condition, he was receiving treatment with Taxotere in late 2011.

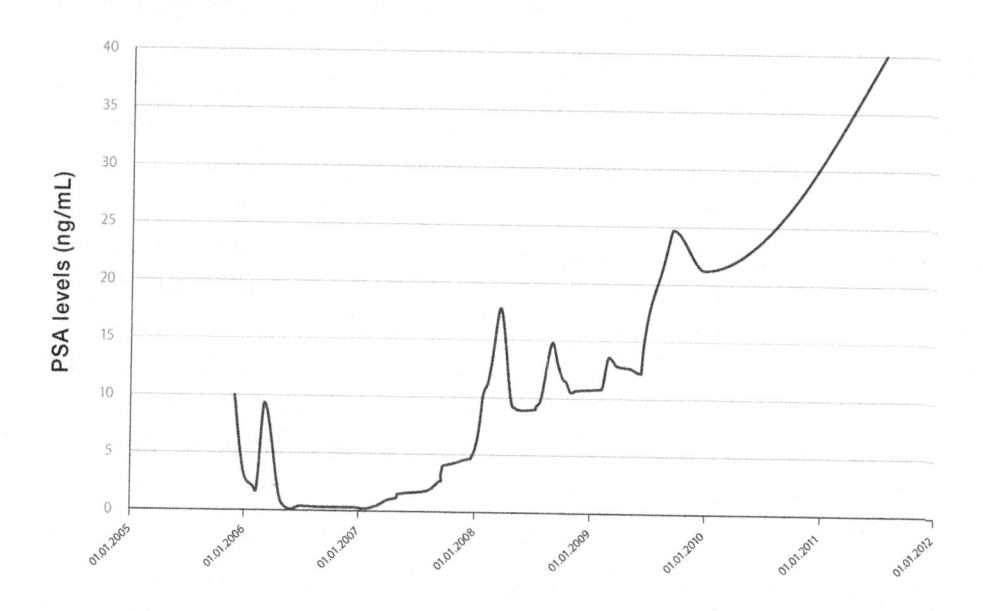

FIGURE 22

The changes in tumor marker levels in patients diagnosed in 2005 with multiple bone metastases and PSA levels at 540 ng/mL

Head and neck tumors

Age of the patient	Sex of the patient	Date of the diagnosis	Start of deuterium depletion	Last update
56 years old	Female	March 1989	December 1992	March 1999

A 56-year-old female patient underwent surgery for a tumor of the tongue (diagnosed a year prior to the surgery) in March 1989. In 1990, the recurrence of cancer was observed, and the patient repeatedly received treatment. Another recurrence was detected in late 1992 when the patient started consuming DDW. Subsequently, the tumor's size started to decrease. A biopsy in March 1993 did not confirm the presence of any tumor tissues. The patient consumed DDW throughout 1993 and had a break from January 1994. A significant progression of cancer was observed, and a repeated surgery took place in August 1994. The patient regularly consumed DDW from June 1994. Another biopsy was performed in February 1997 with positive results. After increasing the dose of DDW, the size of the lumps reduced, and their mass softened. The patient showed significant progression of the disease in spring 1998 when a chest X-ray confirmed a metastasis in the lung. After increasing the dosage, the patient coughed up viscous phlegm. The tumor in her oral cavity softened, the size of the solid tumor next to her esophagus reduced, and the sensitivity of the tumor near her ear decreased. The patient lived for one more year and died in spring 1999. A total of ten years passed since the first surgery, despite the repeated (three times) recurrence of the disease in the first three years when she did not consume DDW.

Age of the patient	Sex of the patient	Date of the diagnosis	Start of deuterium depletion	Last update
63 years old	Female	June 1992	July 1993	December 2002

By June 1993, a 63-year-old female patient underwent surgery three times for recurring oral cavity cancer. Then, as she did not consent to a partial mandibulectomy, she received full dose irradiation. At this point, she started to consume DDW. By August 1993, the lesion under her tongue healed and the tumor on her chin softened. By September, the lump on her neck receded and in October, the tumor was no longer visible. No events of medical importance took place until September 1997, when she had a non-healing wound in her oral cavity. The patient received two cytostatic treatments starting from November 1997. The third one did not take place, as the wound fully healed. The patient continued to consume DDW while enjoying a good quality of life and excellent physical condition. The patient died at the age of seventy-two in December 2002 while leading an active life.

Colorectal cancer

Age of the patient	Sex of the patient	Date of the diagnosis	Start of deuterium depletion	Last update
55 years old	Male	December 2006	January 2007	April 2007

A 55-year-old male patient was diagnosed with an eight-centimeter long, 1.5 cm-thick tumors at the circumference of the colon. Oncology care team conference first suggested radiotherapy for a successful surgery. The patient started to consume DDW before radiotherapy and continued for the duration of radiotherapy. Four months later, complete regression of cancer was confirmed. This observation was in line with previous findings that showed a synergistic effect of DDW and radiotherapy.

Age of the patient	Sex of the patient	Date of the diagnosis	Start of deuterium depletion	Last update
66 years old	Male	March 1998	May 2019	February 2015

A 66-year-old male patient had a family history of polyposis, which significantly increases the risk of colon cancer formation. The first surgery of the patient took place in March 1998, followed by another one in October 2000, and yet another one in April 2001. The patient started to consume DDW from May 2001, after the third surgery. Over the following ten years, he repeated DDW courses thirteen times lasting a total of forty-six months. The disease did not recur after 2001. The last update on the patient's condition reported that the patient did not have any complaints in 2015.

Ovarian cancer

Age of the patient	Sex of the patient	Date of the diagnosis	Start of deuterium depletion	Last update
52 years old	Female	July 1995	April 1996	April 2014

A 52-year-old female patient was diagnosed in summer 1995 with ovarian cancer that was confirmed to be a highly differentiated adenocarcinoma. The uterus and omentum were also affected. Following the surgery, the patient received eight cytostatic treatments (Carboplatin, Cisplatin), which ended in March 1996. The patient consumed DDW for two years after the end of chemotherapy. According to the last update, she was in a good general condition eighteen years after the diagnosis in 2014.

Cervical cancer

Age of the patient	Sex of the patient	Date of the diagnosis	Start of deuterium depletion	Last update
49 years old	Female	November 1993	November 1993	July 1995

A 49-year-old female patient was diagnosed in November 1993 with an inoperable cervical tumor attached to the ovaries. The patient's condition gradually improved following the consumption of DDW (90 ppm) from March 1994. She kept gaining weight and an examination in May did not show any signs of tumor around the cervical opening. A previous intestinal stricture of the patient disappeared by June, and she was in a good general condition. In November 1994, she switched to less-depleted DDW (135 ppm). Four months later, a check-up confirmed the progression of cancer. For a further five months, there were updates on the patient's condition. In this period, she consumed DDW in a dosage lower than recommended and a gradual decline in her condition was reported.

Melanoma malignum

Age of the patient	Sex of the patient	Date of the diagnosis	Start of deuterium depletion	Last update
51 years old	Male	July 1994	November 1994	March 1999

In a 51-year-old male patient, a Clark Level III melanoma originating from a birthmark on the left side of the abdomen was removed in July 1994. In August of the same year, a block dissection was performed in the left armpit. The patient received DTIC and Interferon. The patient consumed DDW from the above date onwards. He was symptom-free in 1995, 1996, and 1997. During this period, only one surgery was performed, when in May 1996, a lymph node was removed that had not been growing and had been there for one and a half years. The microscopic description of the removed lymph node includes the following: "[...] the major part of the mass is occupied separate, but interlinked areas of tumor tissue [...]. Dividing cells are frequently observed. Necrotized areas are observed in the tumor tissue. The tumor is surrounded by a thick capsule. In the examined section, no infiltration of the tumor into the adipose tissue is observed."

With respect to the prolonged period spent without any symptoms, the patient interrupted the consumption of DDW after more than three years, from early 1998. One and a half months later, a lump was observed on the patient's chest. The patient did not continue the consumption of DDW and died in March 1999.

Age of the patient	Sex of the patient	Date of the diagnosis	Start of deuterium depletion	Last update
46 years old	Male	October 1991	October 1994	March 2020

A 46-year-old male patient was operated on with melanoma malignum for the first time in 1991. In April 1992, lymph nodes from the armpit were removed, followed by another surgery in spring 1994. The patient subsequently received DTIC treatment. By the autumn, a metastasis was detected behind one ear and in the liver. He then received treatment with Intron-A. A month after starting the DDW course (in November 1994), a CT scan showed that a previously observed liver metastasis was no longer visible. The size of two metastases reduced and one stagnated. Other examinations (in March, June and December 1996, in February and December 1997, and in February 1998) showed a gradual regression of liver metastases. An excerpt from the summary of the CT scan from April 1999: "Comprehensive examinations do not reveal the previously described minor residual lesions. CT scans show no circumscribed structures in the liver." The patient consumed DDW exclusively with one interruption for eight years, starting from 1994. After that, his daily intake was (and is, at the time of writing, in spring 2020) half a liter. Throughout twenty-five years of consuming DDW, the patient had no symptoms related to the disease.

Liver cancer

Age of the patient	Sex of the patient	Date of the diagnosis	Start of deuterium depletion	Last update
66 years old	Male	June 1985	November 1994	December 1995

A 66-year-old male patient was diagnosed in 1985 with a malignant tumor originating from the liver's connective tissue. By the start of deuterium depletion, a series of surgeries were performed to reduce the tumor's size. Another surgery was performed to cut off the blood flow to the tumor. In May 1994, due to tumor-related stomach bleeding, a partial gastrectomy was performed and bowels were surgically connected with the stomach. The patient experienced a loss of consciousness and abrupt drops in blood sugar levels due to obstructive jaundice. In October 1994, he was admitted to the hospital due to a severe deterioration of his condition. An ultrasound scan in November showed a 17 × 21 cm tumor. After the consumption of DDW, the patient's appetite improved, he gained three kilograms of weight and could move again. In late November, he could leave the hospital. His jaundice improved (as evidenced by testing) and the increased enzyme levels due to the tumor were reduced. In February 1995, an ultrasound scan detected a 16 × 14 cm tumor. After a year of DDW treatment, the jaundice was barely detectable anymore. The improvement of the flow of bile was evidenced by the fact that the previously clay-colored stools became normal again. The patient's condition significantly deteriorated and he experienced a urinary tract obstruction in late November and December. Subsequently, he could not consume DDW any longer and died on December 27, 1995. The significant improvement of the patient's condition starting from November 1994 can only be attributed to DDW, as the patient received no other therapies at that time.

Grade I astrocytoma

Age of the patient	Sex of the patient	Date of the diagnosis	Start of deuterium depletion	Last update
12 years old	Female	April 2000	August 2000	May 2018

A 12-year-old girl was diagnosed with grade I astrocytoma that occupied a large part of the spinal canal. Conventional treatments were not an option for the treatment of the disease. The patient regularly consumed DDW for eight years and repeated courses of DDW over the next six years. An MRI scan in 2010 showed that the extension of the tumor was reduced by two centimeters compared to the scan in 2002. Starting from late 2014, the patient did not consume DDW. Her condition did not necessitate further check-ups. By 2015, the patient graduated from college and was symptom-free in 2018.

Grade III astrocytoma

Age of the patient	Sex of the patient	Date of the diagnosis	Start of deuterium depletion	Last update
29 years old	Male	June 1991	March 1995	June 2001

A 29-year-old male patient was diagnosed in 1991 with a brain tumor following epileptic seizures. The seizures were then managed with medication before surgery in 1994. The patient consumed DDW for forty-four months without interruption (while reducing the concentration from 90 ppm to 62 ppm). During this time, the initially frequent and severe seizures gradually subsided and became infrequent. In 1998, the patient spent weeks without seizures and even the infrequent seizures were not as severe as before. Starting from March 1999, after four years, the concentration of DDW was then increased back to 100 ppm from the previous 62 ppm. In one month, the patient experienced headaches and his condition rapidly deteriorated. An MRI scan confirmed the presence of a fist-sized tumor and cyst. Increasing the dose of DDW could not halt the growth of the tumor. However, the patient underwent successful surgery and could leave the hospital four months later.

The patient's case is a good example of how important the inhibitory effect of DDW has on tumor growth. The dosage was adequate for forty-two months and there was no need to reduce it or increase its deuterium concentration even after four years. The patient relapsed within two months after the second surgery and died in 2001.

Glioblastoma

Age of the patient	Sex of the patient	Date of the diagnosis	Start of deuterium depletion	Last update
44 years old	Male	August 1995	December 1995	November 1997

A 44-year-old male patient underwent surgery in August 1995 for a temporoparietal glioblastoma on the left side. Following the surgery, he received radiotherapy. The patient started consuming DDW in December 1995. Half a year later, in January 1996, a CT scan showed a slight growth of ring-like structures in the surgical area. In April 1996, contrast inhomogeneity was observed with a minor progression of a recurring tumor. In October 1996, a CT scan raised the possibility of a solid tumor recurrence with a diameter of one centimeter. In April 1997, the temporoparietal tumor was approximately three centimeters in diameter and occupied more space. BCNU chemotherapy was started in 1997. The patient died after two years, in November 1997.

Age of the patient	Sex of the patient	Date of the diagnosis	Start of deuterium depletion	Last update
54 years old	Female	September 2014	October 2014	October 2018

A 54-year-old female patient was diagnosed with brain cancer in September 2014, which was identified as glioblastoma. The patient underwent surgery and received radiotherapy and Temozolomide for six weeks. Three weeks after the surgery, but prior to the radiotherapy, the patient began consuming DDW. She consumed 85 ppm DDW for a year, then switched to 65 ppm DDW. According to updates in 2018, all check-ups yielded negative results.

Neurofibromatosis

Age of the patient	Sex of the patient	Date of the diagnosis	Start of deuterium depletion	Last update
12 years old	Female	August 1994	January 1996	November 2007

A 12-year-old girl was receiving treatment for optic nerve glioma caused by neurofibromatosis resulting in amaurosis in both eyes. In August 1994, her underlying condition resulted in a multiplex central nervous system lesion, which manifested in a loss of hearing, facial, oculomotor, and abducens paresis. Also, the lower and upper extremities were paralyzed. Carboplatin/VP-16 chemotherapy was administered and an MRI was used to assess effectiveness. Results showed significant growth and the progression of the disease. Subsequently, the patient did not receive any conventional treatments. The patient started to consume DDW in January 1996. In the beginning, she only consumed Vitaqua (130 ppm), then she switched to 85 ppm DDW. An MRI scan in November 1996 observed a significant reduction in the size of the lump and accumulation of contrast agents. The regression of cancer was identified. The patient's speech improved continuously following the consumption of DDW. Her mobility also gradually improved from autumn 1996. In January 1997, she could walk without support. In August 1997, the tumor reduced to half the size of November in the previous year. MRI scans did not show a progression of the disease in December 1998 either. The patient was later homeschooled and according to the last update in 2007, was in a good general condition.

Bone marrow cancer

Myeloma

Age of the patient	Sex of the patient	Date of the diagnosis	Start of deuterium depletion	Last update
65 years old	Male	September 1994	October 1994	December 1998

A 65-year-old male patient was diagnosed with myeloma multiplex in September 1994. Prior to this, he had had a pathological rib fracture. The patient received cytostatic treatment two months before the DDW course. By January 1995, after two months of consuming DDW, the patient gained five kilograms. In March, serum protein electrophoresis testing confirmed significant improvement. Two months later, no pathological changes were detected. The patient received the last of a series of treatments in July. In September 1995, the serum levels were in the normal range according to electrophoresis testing and blood counts (no pathological changes were detected for the latter one even in 1996). In April, bone scintigraphy also did not show any of the previously accumulated structures and the check-up in September was negative. The patient spent time in a thermal bath in October 1996. By the end of his stay, he experienced discomfort and pain between the ribs. Another treatment was started in November 1996. A tumor of significant size formed on the sternum by March 1997. The tumor regressed by May and recurred in August. In the meantime, metastases were observed in the thoracic intervertebral spaces. The patient consumed DDW with minor interruptions during the conventional treatments and lived for two more years. He died four and a half years after the diagnosis. It is reasonable to conclude that his condition began to deteriorate during the time he spent at the thermal bath.

Acute myeloid leukemia (AML)

Age of the patient	Sex of the patient	Date of the diagnosis	Start of deuterium depletion	Last update
26 years old	Male	June 1994	January 1995	December 2008

In a 26-year-old male patient, the swelling of lymph nodes was observed after being sick and having a fever in September 1992. With respect to the histologic diagnosis (Hodgkin's lymphoma), the patient was administered ABDV, and, from February 1994, COPP therapy. In June 1994, a bone marrow biopsy confirmed AML M4. The patient refused to continue chemotherapy after three rounds of treatment in October 1994. Blood counts from January 3 and January 10, 1995, show an increase of blastoid cells in peripheral blood. The patient started consuming DDW on January 10, 1995, and blood counts on January 31 did not show the presence of blastoid cells. The patient consumed DDW continuously until March 1997, with all check-ups yielding negative results. Starting from January 1998, he repeatedly consumed DDW for a few months for safety's sake. In 2008, sixteen years after the diagnosis, the patient was in a good general condition.

Age of the patient	Sex of the patient	Date of the diagnosis	Start of deuterium depletion	Last update
32 years old	Female	November 2006	December 2006	March 2020

A 32-year-old female patient was diagnosed with AML in November 2006. Bone marrow testing confirmed a 60% infiltration. The patient started consuming DDW a week after the diagnosis, while also receiving chemotherapy at the same time. Conventional therapies took place between November 2006 and January 2007. The patient declined further cytostatic treatments after this. As a result of deuterium depletion and chemotherapy, no blastoid cells were detected five weeks after the diagnosis. A bone marrow test four months later did not confirm the patient's underlying condition. The patient consumed DDW at first for nine months, gradually decreasing deuterium concentration from 85 ppm to 65 ppm, then finally to 45 ppm. Subsequently, after a four-week interruption, she consumed DDW for another four months, while reducing the deuterium concentration from 105 ppm every fourth week, ultimately down to 25 ppm. Since no recurrence occurred during the interruption or application of DDW, the duration of the interruption was extended to two, four, six, and twelve months, respectively. The patient had three- to four-month courses of DDW until early 2011. According to the last updates on her condition, the patient was in a good general condition fourteen years after the diagnosis.

Chronic lymphocytic leukemia (CLL)

Age of the patient	Sex of the patient	Date of the diagnosis	Start of deuterium depletion	Last update
64 years old	Male	October 1995	January 1996	October 1999

A 64-year-old male patient was diagnosed with B-cell CLL in 1992. Starting from late 1995, due to the patient's underlying condition, his white blood cell count increased. Increasingly severe anemia and thrombocytopenia were observed. The conditions were treated with red cell and platelet concentrates. A CT scan showed shadows on the apex of both lungs. Multiple lymph nodes in the size of 1.5–2 cm were visible in the mediastinum. The scan showed an enlargement of the liver and the spleen with a partial confluence of lymph nodes in the mesenterium. The disease progressed despite the conventional treatment. The patient was weak and bedridden, continuously losing weight. He consumed DDW from January 1996. Then, in the following two months, he needed blood transfusions on fewer occasions. Later, he did not need them anymore based on his blood counts. His general condition improved and four months later, the enlarged lymph nodes in the neck were no longer palpable. By the end of the year, the patient experienced complaints in the lower abdomen, which was followed by surgery in January 1997 (several palpable lymph nodes were removed). In spring 1998, the patient was again diagnosed with enlarged axillary lymph nodes, after which he consumed DDW for three months. A year later, by May 1999, the patient had gained eighty kilograms in body weight, the enlarged lymph node was not palpable, and he enjoyed good general health. The patient died in late 1999 of pneumonia.

Age of the patient	Sex of the patient	Date of the diagnosis	Start of deuterium depletion	Last update
41 years old	Male	February 2006	February 2006	March 2020

A 41-year-old male patient was diagnosed with CLL in February 2006. Apart from a high number of white blood cells (16,000), the patient had a sizable lymph node in the neck and his spleen was twice the normal size. A nine-centimeter lymph node conglomerate in the abdominal region was revealed by ultrasound scan. The patient started consuming DDW immediately after the diagnosis. His white blood cell counts went back to normal (dropped below 10,000) and the size of the lymph node also reduced significantly. In the beginning, the patient consumed DDW for more than three years. During this period, the size of his spleen went back into the normal range. The size of conglomerate lymph nodes in the abdominal region decreased, and the lymph node in the neck almost completely receded. Upon the interruption of deuterium depletion, a slight progression of cancer was recorded, which was successfully reversed using a DDW course. After the first several years of continuous DDW consumption, the patient consumed DDW twelve times over fourteen years in a few monthly courses. Until the time of writing, he has not needed chemotherapy, and he is free of symptoms and complaints.

CHAPTER FOURTEEN

Advice on Establishing the Dosage

Recommendation for Healthy People, Prevention, Enhancing Performance

This group comprises healthy people who do not have and have never had any cancer. If we take a look at the cancer cases by age, it is apparent that there is a surge in cancer cases after age 40 (see Fig. 1). In terms of the preventive use of deuterium depletion, the healthy population can be broken down into two more groups, taking into account age and other risk factors. Accordingly, two recommendations are made for the use of DDW. For each DDW protocol, four parameters can be used to improve effectiveness. The four parameters are the following:

a) The concentration of deuterium in DDW

b) The duration of DDW consumption

c) The daily intake of DDW

d) The frequency of repeating a DDW course (this is of utmost importance)

This is the basic distinction between the two recommendations.

H/1 Protocol

This protocol refers to a population of individuals under forty or fifty, with low risks and no demonstrable genetic predisposition (risk), family history, unhealthy lifestyle factors (smoking, obesity, occupational risks, etc.).

This group has the lowest risk for the formation of cancer. People who do not smoke, have a healthy diet and normal body weight, are not exposed to carcinogenic substances and harmful radiation with no confirmed family history of cancer belong to this group. For people in this group, we recommend repeating the DDW regimen every two to three years for three to four months. Recommended deuterium concentration: 125 or 105 ppm.

Recommended deuterium concentration	Recommended daily intake	Recommended duration of the cure
125 or 105 ppm	1.5–2.0 liters	3–4 months

H/2 Protocol

This protocol refers to people over fifty years of age and/or people who, regardless of age, are in a higher risk group for one or more factors.

Such patients are those who possess a genetic predisposition to cancer, live in polluted urban or industrial areas, are exposed to occupational risks, or are at risk due to their lifestyle (smoking, alcohol, unhealthy diet, obesity, stress, etc.). These patients are at a greater risk of developing cancer. In this group, it is recommended to repeat the DDW treatment every one to two years for four months. Recommended deuterium concentration: 125 or 105 ppm.

Recommended deuterium concentration	Recommended daily intake	Recommended duration of the cure
125 or 105 ppm	1.5–2.0 liters	4 months

Recommendations for people who are not yet diagnosed with cancer, but are examined for the suspicion of cancer

P/D Protocol

This group comprises individuals whose symptoms raised the suspicion of cancer and are currently undergoing diagnostic examinations. The use of deuterium depletion during the examinations offers an immediate and safe preventive option before starting conventional treatment if the disease is later confirmed. In this group, DDW treatment is recommended for up to four months in the case of a negative finding, or longer depending on the stage of cancer and conventional treatments used in the case of a cancer diagnosis. Recommended deuterium concentration: 105 or 85 ppm.

Recommended deuterium concentration	Recommended daily intake	Recommended duration of the cure
105 or 85 ppm	1.5–2.0 liters	until an accurate diagnosis is established and with conventional treatment, or four months in case of negative findings

Recommendations for cancer patients

Effective oncological treatment must serve two main purposes: (1) It must ensure that the patient becomes macroscopically cancer-free, (2) This condition must be maintained for the longest time possible. A major cause of cancer-related deaths is that none of the currently available tools of oncology corresponds to these expectations. In an ideal case, it is possible to achieve a cancer-free condition. However, the rate of recurrence is still very high, and we cannot ensure that patients become completely cancer-free. Below are some recommendations that may help to give a proper answer to the multi-faceted challenges arising from the stages and diversity of cancer, and the possible combinations of conventional treatments. Integrating deuterium depletion into the treatment regimen may also help achieve a cancer-free condition irrespectively of which stage of cancer a patient is in, and to maintain that condition for as long as possible. Further clinical research and trials are needed to achieve this.

Recommendations for Patients Who Have Recovered to Prevent the Relapse of Cancer

Patients who have undergone successful treatments and are cancer-free can be broken down into two groups. The first group comprises those patients whose treatment regimen did not include deuterium depletion. The second group consists of patients who received DDW to achieve a cancer-free condition. The first group can be broken down into two other subgroups by when patients achieved a cancer-free condition, either recently or years ago. With respect to the above considerations, the recommendations are as detailed below.

Recommendations for Patients Who Have Achieved a Cancer-Free Condition with Conventional Therapies to Prevent Disease Recurrence

C/R/1 Protocol

Patients in this group have been cancer-free for one to two years thanks to conventional treatments, the disease has not recurred since diagnosis and treatment, and patients are symptom-free.

Within the group, further subgroups may be formed depending on the risk of recurrence for a specific patient. It should be considered, however, that it is almost impossible to determine this risk with absolute certainty, and the recommendations are based on the worst-case scenario of a recurrence. Observing such recommendations may seem like overkill for a patient with a good prognosis, but the most important goal to consider is a full recovery. Recommended deuterium concentrations: 105, 85, and 65 ppm.

Recommended deuterium concentration	Recommended daily intake	Recommended duration of the cure
105 ppm	1.5–2.0 liters	1.5–2 months
85 ppm	1.5–2.0 liters	1.5–2 months
65 ppm	1.5–2.0 liters	1–2 months

2-4 month pause

Recommended deuterium concentration	Recommended daily intake	Recommended duration of the cure
105 ppm	1.5–2.0 liters	1.5–2 months
85 ppm	1.5–2.0 liters	1.5–2 months
65 ppm	1.5–2.0 liters	1–2 months

5–6 month pause

Recommended deuterium concentration	Recommended daily intake	Recommended duration of the cure
105 ppm	1.5–2.0 liters	1.5–2 months
85 ppm	1.5–2.0 liters	1.5–2 months
65 ppm	1.5–2.0 liters	1–2 months

8–10 month pause

Recommended deuterium concentration	Recommended daily intake	Recommended duration of the cure
105 ppm	1.5–2.0 liters	2 months
85 ppm	1.5–2.0 liters	2 months

Repeat the two-month DDW-105 and two-month DDW-85 courses annually for two to three years.

C/R/2 Protocol

Patients in this group have been cancer-free for at least three to four years since the end of conventional treatments. Since the diagnosis and successful treatment of the disease, there has been no recurrence and patients are symptom-free. Recommended deuterium concentration: 105, 85, and 65 ppm.

Recommended deuterium concentration	Recommended daily intake	Recommended duration of the cure
105 ppm	1.5–2.0 liters	1.5–2 months
85 ppm	1.5–2.0 liters	1.5–2 months
65 ppm	1.5–2.0 liters	1–2 months

10–12 month pause

Recommended deuterium concentration	Recommended daily intake	Recommended duration of the cure
105 ppm	1.5–2.0 liters	2 months
85 ppm	1.5–2.0 liters	2 months

Repeat the two-month DDW-105 and two-month DDW-85 courses every year for another two to three years.

Recommendations for patients who have become cancer-free during the complementary use of deuterium depletion to prevent a relapse

C/R/3 Protocol

This group includes patients who have used deuterium depletion in addition to conventional treatments and have already reduced their deuterium concentration in several steps. Given the variety of conventional treatments, their combinations, and the different sensitivities of cancer types, a general recommendation is made. Details and other considerations are described alongside the combination of conventional treatments with DDW. After achieving a cancer-free condition, it is recommended to follow this C/R/3 Protocol.

Recommended deuterium concentration	Recommended daily intake	Recommended duration of the cure
Deuterium levels at the time of the remission	1.5–2.0 liters	2–3 months
20 ppm lower D-concentration	1.5–2.0 liters	2 months
20 ppm more lower D-concentration	1.5–2.0 liters	2 months

If it is not possible to further reduce the deuterium concentration at the time of remission, it is recommended to continue deuterium depletion at the lowest level for four to six months, and then to use the C/R/1 protocol after an interruption of two to three months.

Recommendations for Cancer Patients to Achieve a Cancer-Free Condition, Taking into Account the Conventional Treatments Used

The patient is about to undergo surgery

C/C/Op Protocol

This group comprises patients with the indication of a surgical intervention at the time of the diagnosis. Recommended deuterium concentration: 105 or 85 ppm.

Recommended deuterium concentration	Recommended daily intake	Recommended duration of the cure
105 ppm	1.5–2.0 liters	Up until the day of the surgery
Water with normal deuterium levels	1.5–2.0 liters	5–7 days after the surgery
105 or 85 ppm	1.5–2.0 liters	2 months, with the deuterium concentration applied before the surgery

Following the C/R/1 Protocol is recommended after successful surgery and while being in a cancer-free condition. Recommended deuterium concentration: 105 or 85 ppm.

Inoperable patients receiving chemotherapy

I/C/C/Chem Protocol

Many patients are inoperable at the time of diagnosis. One option in these cases is to provide pre-treatment applying chemotherapy, which aims to achieve operability. Following the below protocol is recommended. Deuterium depletion may be started with a deuterium concentration of 105 or 85 ppm.

Recommended deuterium concentration	Recommended daily intake	Recommended duration of the cure
105 or 85 ppm	1.5–2.0 liters	for 2–3 months until the first check-ups
105 or 85 ppm	1.5–2.0 liters	for a further 2–3 months, if follow-up examinations confirm the regression of cancer
85 or 65 ppm	1.5–2.0 liters	2–3 months, if follow-up examinations do not confirm any improvement, but the treatment is still ongoing
85 or 65 ppm	1.5–2.0 liters	2–3 months, if follow-up medical checkups indicate a continuous regression or the patient's condition is stable
65 or 45 ppm	1.5–2.0 liters	2–3 months, if follow-up medical checkups indicate a continuous regression or the patient's condition is stable
45 ppm	1.5–2.0 liters	2–3 months, if follow-up medical checkups indicate a continuous regression or the patient's condition is stable
25 ppm	1.5–2.0 liters	2–3 months, if follow-up medical checkups indicate a continuous regression or the patient's condition is stable

After ten to twelve months of continuous consumption of DDW, an interruption of one to two months is recommended (if a cancer-free condition is not yet achieved). Following the interruption, another course with a deuterium concentration of 105 or 85 ppm may be started.

Patients receive aftercare with adjuvant chemotherapy following a successful surgery.

2/C/C/Chem Protocol

For some patients who have undergone successful surgery, adjuvant chemotherapy is used to increase the duration of a progression-free period and prevent the recurrence of cancer. In some cases, the circumstances of surgery, the type, size, and pathology of cancer offer hope for the patients of a full recovery. In these cases, where deuterium depletion is not used in the entire or some of the duration of chemotherapy, DDW is only used after or in the last phase of the treatment. If the prognosis is still poor despite successful surgery, the below protocol differs only in recommending DDW-105 for the entire time of chemotherapy.

Recommended deuterium concentration	Recommended daily intake	Recommended duration of the cure
Normal	1.5–2.0 liters	During the entire time of chemotherapy, or until the last 1–2 treatments
105 ppm	1.5–2.0 liters	After completing chemotherapy, or simultaneously with the last 1–2 treatments
105 ppm	1.5–2.0 liters	For 2 months, after the last chemotherapy treatment
85 ppm	1.5–2.0 liters	2 months
65 ppm	1.5–2.0 liters	2 months

3–4 month pause

Recommended deuterium concentration	Recommended daily intake	Recommended duration of the cure
105 ppm	1.5–2.0 liters	2 months
85 ppm	1.5–2.0 liters	2 months
65 ppm	1.5–2.0 liters	1.5–2 months

5–6 months pause

Recommended deuterium concentration	Recommended daily intake	Recommended duration of the cure
105 ppm	1.5–2.0 liters	2 months
85 ppm	1.5–2.0 liters	2 months
65 ppm	1.5–2.0 liters	1.5–2 months

8–10 months pause

Recommended deuterium concentration	Recommended daily intake	Recommended duration of the cure
105 ppm	1.5–2.0 liters	2 months
85 ppm	1.5–2.0 liters	2 months

Repeating a two-month DDW-105 course and a two-month DDW-85 course every year, for two to three years.

Recommendations for patients with glioblastoma, fitted to the Stupp protocol

3/C/C/Chem Protocol

Patients usually receive radiotherapy and are administered Temozolomide, but for a significant proportion of these patients, the surgical removal of the tumor is not an option. The below protocol contains recommendations for such patients:

Recommended deuterium concentration	Recommended daily intake	Recommended duration of the cure
85 ppm	1.5–2.0 liters	Radiotherapy for 6 weeks simultaneously with Temozolomide, for 1–2 weeks after the last radiotherapy treatment
65 ppm	1.5–2.0 liters	2–3 months
45 ppm	1.5–2.0 liters	2–3 months
25 ppm	1.5–2.0 liters	5–6 months
Normal	1.5–2.0 liters	1–3 months
85 ppm	1.5–2.0 liters	Follow-up examinations confirm an ongoing regression, or if the patient's condition stagnates, reduce the concentration by 20 ppm every 2 months. If the disease progresses, then reduce the concentration by 20 ppm every month

Patients are inoperable and receive hormone treatment

C/C/Horm Protocol

Recommended deuterium concentration	Recommended daily intake	Recommended duration of the cure
105 or 85 ppm	1.5–2.0 liters	For 2–3 months during hormone therapy until the first follow-up examinations
105 or 85 ppm	1.5–2.0 liters	For a further 2–3 months, if follow-up examinations confirm regression of cancer
85 or 65 ppm	1.5–2.0 liters	For 2–3 months, if follow-up examinations do not confirm regression of cancer, the treatment is continued
85 or 65 ppm	1.5–2.0 liters	For 2–3 months, if follow-up examinations confirm a continuous regression of cancer, or if the disease stagnates
65 or 45 ppm	1.5–2.0 liters	For 2–3 months, if follow-up examinations confirm a continuous regression of cancer, or if the disease stagnates
normal	1.5–2.0 liters	For 2–3 months, if the follow-up examinations confirm tumor marker levels shifting to the normal range for 3 months
105 ppm	1.5–2.0 liters	For 2–3 months, if follow-up examinations confirm that the patient is in a stable condition without the progression of cancer for 2–3 months during the interruption of DDW consumption; hormone therapy may be interrupted at the same time when starting another DDW course
85 ppm	1.5–2.0 liters	2–3 months, if follow-up medical checkups indicate no change
65 ppm	1.5–2.0 liters	2–3 months, if follow-up medical checkups indicate no change
Normal	1.5–2.0 liters	For 2–3 months, if follow-up examinations confirm tumor marker levels to be within the normal range, or extend the period of interruption by 2–3 months, or halt hormone therapy, or resume the use of DDW for a period without hormone therapy

Recommended deuterium concentration	Recommended daily intake	Recommended duration of the cure
105 ppm	1.5–2.0 liters	For 2–3 months, if follow-up examinations confirm that the patient is in a stable condition without progression of cancer, hormone therapy may be interrupted at the same time as starting a new DDW course
85 ppm	1.5–2.0 liters	2–3 months, if follow-up medical checkups indicate no change
65 ppm	1.5–2.0 liters	2–3 months, if follow-up medical checkups indicate no change

Optimally, it is recommended to repeat the above cycles. The complementary use of DDW enables a temporary interruption of hormone therapy, preventing or mitigating the onset of resistance to hormones. Temporarily halting the use of DDW prevents the onset of resistance to DDW. Hormone treatment may be restarted to prevent progression while DDW is stopped or, if the patient's condition allows, they may be exempted from treatment. Another cycle using deuterium depletion may be started after two to three months of interruption.

Patients receive radiotherapy

C/C/Radther Protocol

In some of the newly diagnosed, inoperable cases, radiotherapy is used to make the patient operable following the regression of cancer. Using the below protocol may significantly enhance the effectiveness of radiotherapy and increase the chance of reaching an operable stage.

Recommended deuterium concentration	Recommended daily intake	Recommended duration of the cure
105 ppm*	1.5–2.0 liters	Before radiotherapy, if possible, then during the entire time of radiotherapy
105 ppm	1.5–2.0 liters	For 4–5 weeks for radiotherapy with a duration of 5–6 weeks, following the last irradiation, if radiotherapy took a shorter time, the for a proportionally shorter period after the last irradiation
85 ppm	1.5–2.0 liters	2–3 months
65 ppm	1.5–2.0 liters	2 months

* For glioblastoma, the treatment begins at 85 ppm.

Continue the treatment with the C/R/1 protocol following successful surgery and achieving a cancer-free condition. This does not apply to patients who have undergone surgery for glioblastoma (see the next chapter, "*Special advice on establishing the dosage*").

Special advice

Stomach cancer is a type of cancer that responds to treatment with deuterium depletion even in late stages. The main reason for this is that consuming DDW may help achieve a significant reduction of the local deuterium concentration in the tumor's surroundings. This means that depending on the D concentration, for example when DDW-105 is consumed, there will be a concentration difference of forty to forty-five ppm between the water content of the tumor cells and the D concentration of DDW. This significant difference may result in a rapid and powerful impact, which may also cause minor bleeding. When treating stomach cancer, stools and blood counts should therefore be monitored so that early intervention may be made if major blood loss occurs. For this type of cancer, in case of the rapid response of the tumor, it is recommended to reduce the dose and elevate the deuterium concentration, reverting to 125 ppm from 105 ppm.

For this type of cancer, special advice is that DDW should be consumed in small sips throughout the day, in addition to one-glass portions, to ensure that fresh, low-deuterium water is constantly present in the stomach.

A similar direct relationship and effect may be established for oral cavity cancers. It is recommended that the patient keep DDW in the mouth for a few minutes before ingesting it. This is to ensure that the reduction of local deuterium concentration takes effect for a longer period and that DDW remains in contact with the tumor-affected area.

The case of a 46-year-old male patient diagnosed in 1991 with melanoma and liver metastases was described in detail in the chapter, *"A demonstration of the effectiveness of deuterium depletion through case studies."* In the *Special advice on establishing the dosage*, it is highlighted again. The reason for this is that at the time of writing, the patient had been consuming DDW without interruption for twenty-four years. In that time, the metastases completely regressed and the patient has had no complaints whatsoever related to his underlying condition. A contrary example is that of a patient who underwent surgery for melanoma in 1995 and had no confirmed metastases. The patient consumed DDW for two and a half more years and never repeated the courses. Metastases were confirmed six years later in the spleen and the liver. Treatments could not halt the progression of the disease and the patient died half a year later. Melanoma is a type of cancer where repeating low-deuterium DDW courses for several years is crucial.

For cancers of the central nervous system, deuterium depletion may have an important role in preparing the patient for surgery. When operating on brain tumors, doctors often face the problem of even though it is evident that a larger area should be removed to eliminate the tumor, such a surgical intervention would affect crucially important regions. The damage to these vital regions may risk the patient's life and surgery could only be conducted at the cost of substantial and permanent damage. For other cancers, consuming DDW even for a few weeks may significantly enhance operability. For tumors of the central nervous system (provided that the patient's condition allows it), consuming DDW for even a few months is recommended under continuous medical supervision. Doing so may enable surgeons to remove a more compact tumor that is separated from its surroundings. This reduces the residual tumor mass and minimizes the area to be removed. The combined use of DDW and conventional aftercare options help prevent the recurrence of cancer.

The type of central nervous system cancer with one of the worst prognoses is glioblastoma. A mere few months without the progression of the disease is the best result for this type of cancer, and it may be achieved with surgery, radiotherapy, and medication therapy. Evaluating data from patients diagnosed with glioblastoma who received deuterium depletion, it is clear that the life expectancy of this group of patients was three times that of patients receiving conventional treatment alone, with the potential for further improvements in efficacy.

CHAPTER FIFTEEN

Recommendations for Patients Diagnosed with Metabolic Disorders (M Protocols)

In the initial phases of research, a surprising finding was observing a diabetic patient's blood sugar levels decrease as a result of a DDW course (applied due to cancer). This finding inspired us to engage in animal testing. Testing with rats confirmed that the change in deuterium concentrations, coupled with the effects of insulin, has a significant influence on metabolism. A key finding of the experiments was that contrary to experience with cancer, it was not the lowest deuterium concentration that caused the most significant reduction in blood sugar levels. Conversely, it was the 125–135 ppm deuterium concentration, which is only 15–25 ppm lower than natural deuterium levels. After successful preclinical trials, it was confirmed in Phase II clinical trials that triggering a 15–25 ppm reduction in the body's deuterium levels caused insulin resistance in 30% of the patients. In 50% of the patients, a significant decrease in insulin concentration was detected, showing a positive correlation with blood sugar levels [44, 63, 74].

M Protocol

Recommended deuterium concentration	Recommended daily intake	Recommended duration of the cure
125 or 105 ppm	1.5–2.0 liters	3–4 months
125 or 105 ppm	0.5–1.0 liters	Continuous
Normal	1.5–2.0 liters	Until the parameters start deteriorating again (blood sugar levels) when the previous course may be repeated

CHAPTER SIXTEEN

Recommendations for Athletes and Healthy People, to Enhance Physical Performance

Feedback from healthy people consuming 125 or 105 ppm shows improvement in their physical endurance. They were capable of longer physical activity and recovered faster following physical exertion. This experience was used to launch a sports medicine study aiming to investigate how deuterium depletion affects physical performance and the metabolic processes (for example, the metabolism of sugar and pH balance) during physical activity. In the study, five boat-racing athletes consumed normal water, while seven others consumed Preventa 105 ppm deuterium-depleted water for forty-four days. Before the study, all athletes were subjected to performance tests, which were repeated forty-four days later, at the end of the study. The performance test involved running 4 × 1500 meters (T1-T4) with increasing intensity, with two-minute stops in between. The blood's levels of glucose, cation, anion, lactic acid, and acid-base parameters were determined from capillary blood before, after, and five minutes after the end of the performance test. The results were published in the Hungarian Review of Sports Medicine (Sportorvosi Szemle) [75]. The most important conclusions are detailed below.

Under physical exertion, the increased energy intake of muscles is covered with sugar molecules from the bloodstream. Thus, at the start of the performance test, the blood levels of all athletes dropped by twenty-five to thirty-four percent in the points T1 and T2. This decrease was reduced to five to seven percent in the Preventa group after forty-four days. In absolute terms, the initial 1.9–2.6 mmol/liter decrease in glucose concentration in the treatment group was reduced to a concentration change of 0.4–0.5 mmol/liter by the end of the study. The body, sensing a drop in blood glucose levels, immediately reacts to the change and mobilizes glycogen reserves from the liver to increase the concentration of glucose in the bloodstream. This is the reason why after the fourth 1500 meters, and especially at the end of the five-minute break, blood glucose levels exceeded the pre-test levels. Athletes in the Preventa group had a nine percent increase on day zero of the study, which was forty-six percent higher than the values when resting forty-four days later at the end of the study, demonstrating that DDW consumption improved the mobilization of glucose in athletes.

Similarly, significant differences were observed between the lactic acid levels in the control group and the treatment group. The concentration of lactic acid was significantly lower than that of the control group (1.44 mmol/liter vs. 2.54 mmol/liter) following the forty-four-day period of consuming DDW. Even more important, the levels remained significantly low at the time of the first three measurements (T1–T3) (T1: 1.54–2.52, T2: 2.62–4.93, T3: 6.29–8.91 mmol/liter). This suggests that the cells were either deprived of oxygen later during the performance test or that their ability to eliminate lactic acid improved.

Acidification as a result of physical exertion (metabolic acidosis) results in the widening of the so-called serum anion gap. It means that the approximately 10 mmol/liter concentration difference between the serum's positively and the negatively charged ions (Na^+, K^+, Ca^{2+}, Mg^{2+}, and HCO_3^-, Cl^-) increases. The reason for this is an increased lactate concentration. In the study, the anion gap at T1–T3 in the Preventa group consuming DDW was significantly narrower than in the other group. This finding confirms that cells can better compensate for the metabolic changes with a low deuterium concentration, improving the oxygenation of the tissues.

The above data explain why people consuming deuterium-depleted water feel more energetic. The findings also confirm experimental results which indicate that deuterium depletion induces beneficial processes in the body, and these processes are evident during physical activity.

It is important to note that the 10–20 ppm decrease in deuterium concentration as a result of using 105 ppm DDW can provide an optimal deuterium level for the body, thereby improving its functions. It was discussed earlier that if carbohydrates are the prevalent energy source, they maintain the body's deuterium concentration at approximately 150 ppm. This is the result of the fact that over the last fifty to sixty years, nutrition science has made animal fats responsible for the development of cardiovascular diseases, and increased carbohydrate consumption has been recommended as a remedy for this problem, alongside reduced fat intake. As a "result," the number of diabetic patients worldwide is approximately 300 million. In developed countries, the majority of the population is overweight or obese.

For healthy people, the consumption of DDW with concentrations of 125 or 105 ppm and dietary changes are recommended to improve physical performance. Dietary changes mean a reduced carbohydrate and increased fat intake, providing the body with the deuterium levels for optimal functioning.

Appendix

SUMMARY TABLE OF THE DEUTERIUM CONCENTRATIONS IN DIFFERENT NUTRIENTS

Coconut water	156 ppm	Cottage cheese	136 ppm
Soluble corn fiber	155 ppm	Cocoa butter	132 ppm
All-purpose flour	150 ppm	Peanut butter	131 ppm
Egg	146 ppm	Olive oil	130 ppm
Beet sugar	146 ppm	Sunflower seed oil	130 ppm
Corn	145 ppm	Preventa–125	125 ppm
Sorghum	144 ppm	Petroselinic acid	125 ppm
Potato	143 ppm	Butter made from cow milk	124 ppm
Cabbage	143 ppm	Cow fat	121 ppm
Wheat	142 ppm	Pig fat	118 ppm
Carrot	142 ppm	Palm kernel oil	117 ppm
Oat	141 ppm	Preventa–105	105 ppm
Red beet	138 ppm	Preventa–85	85 ppm
Pork meat	138 ppm	Preventa–65	65 ppm
Beef meat	138 ppm	Preventa–45	45 ppm
Chicken meat	137 ppm	Preventa–25	25 ppm
Sodium caseinate	137 ppm		
Spinach	136 ppm		

Bibliography

1. J. M. Bishop, "Cancer: the rise of the genetic paradigm", *Genes and Development,* 9, p. 1309, 1995.
2. T. N. Seyfied, Cancer as a Metabolic Diseases, John Wiley and Sons, 2012.
3. O. Warburg, The Metabolism of Tumors, New York: Richard R. Smith, 1930.
4. A. Bruce, B. Dennis, L. Julian, R. Martin, R. Keith and D. W. James, Molecular Biology of the Cell, New York: Garland Publishing Inc., 1989.
5. Y. Yang, A. N. Lane, C. J. Ricketts, C. Sourbier, M–-H. Wei, B. Shuch, L. Pike, M. Wu, T. A. Roualt, G. L. Boros, T. W. Fan and W. M. Linehan, "Metabolic Reprogramming for Producing Energy and Reducing Power in Fumarate Hydratase Null Cells from Hereditary Leiomyomatosis Renal Cell Carcinoma", *PLOS One,* 8, p. 72179, 2013.
6. D. S. Wishart, "Is Cancer a Genetic Disease or a Metabolic Disease?", *EBioMedicine,* 2, pp. 478–479, 2015.
7. WHO, "https://www.who.int/health-topics/cancer#tab=tab_1" 2018.
8. T. Beardsley, "A war not won", *Scientific American,* 130, pp. 118-126, 1994.
9. American Cancer Society, American Cancer Society Cancer Facts and Figures 2018, 2018.
10. A. Szent-Györgyi, "The living state of cancer", *Physiol Chem Phys,* 12, pp. 99-110, 1980.
11. A. Szent-Györgyi, "The Living State and Cancer. In: Submolecular Biology and Cancer" in *Ciba Foundation Symposium 67,* 1979.
12. G. Somlyai, G. Jancsó, G. Jákli, K. Vass, B. Barna, V. Lakics and T. Gaál, "Naturally occurring deuterium is essential for the normal growth rate of cells", *FEBS Lett.,* 317, pp. 1-4, 1993.
13. T. Berkényi, G. Somlyai, G. Jákli and G. Jancsó, "Csökkentett deutérium-tartalmú (Dd-víz) alkalmazása az állatgyógyászatban", *Kisállatorvoslás,* 3, pp. 114–115, 1996.

14. G. Somlyai, "Biologische Auswirkungen von Wasser mit vermindertem Deuteriumgehalt. Acta medica empirica", *Acta medica empirica,* 7, pp. 381-388, 1997.
15. G. Somlyai, G. Laskay, T. Berkényi, Z. Galbács, G. Galbács, S. A. Kiss, G. Jákli and G. Jancsó, "The Biological Effects of Deuterium-Depleted Water, a Possible New Tool in Cancer Therapy", *Z. Onkol. J. of Oncol.,* 30, p. 4, 1998.
16. G. Somlyai, G. Laskay, T. Berkényi, G. Jákli and G. Jancsó, "Naturally occuring deuterium may have a central role in cell signalling" in *Synthesis and Application of Isotopically Labelled Compounds,* J. R. Heys and D. G. Mellilo, szerk., John Wiley & Sons Ltd., 1998, pp. 137–141.
17. G. Laskay, G. Somlyai, G. Jancsó and G. Jákli, "Reduced deuterium concentration of water stimulates O2-uptake and electrogenic H^+-efflux in the aquatic macrophyte Elodea Canadensis", *Japanese Journal of Deuterium Sciences,* 10, pp. 17-23, 2001.
18. G. Somlyai, G. Jancsó, G. Jákli, T. Berkényi, Z. Gyöngyi and I. Ember, "The Biological Effect of Deuterium Depleted Water, a Possible New Tool in Cancer Therapy", *Anticancer Research,* 21, p. 1617, 2001.
19. M. Szabó, Z. Sápi, T. Berkényi and G. Somlyai, "A deutérium-megvonás hatása állati tumorokra and azok pathológiás képére", *Az állatorvos,* III, 7–8, pp. 22–23,26–27, 2003.
20. G. Somlyai, "A természetben megtalálható deutérium biológiai jelentősége: a deutériumdepletio daganatellenes hatása", *Orvosi Hetilap,* 151, pp. 1455-1460, 2010.
21. G. L. Boros, P. D. Dominic, E. K. Howard, P. R. Justin, J. M. Emmanuelle and G. Somlyai, "Submolecular regulation of cell transformation by deuterium depleting water exchange reactions in the tricarboxylic acid substrate cycle", *Medical Hypotheses,* 87, pp. 69-74, 2016.
22. G. Somlyai, T. C. Que, J. M. Emmanuelle, H. Patel, P. D. Dominic and G. L. Boros, "Structural homologies between phenformin, lipitor and gleevec aim the same metabolic oncotarget in leukemia and melanoma", *Oncotarget,* 25, pp. 50187-50192, 2017.

23. L. G. Boros, G. Somlyai, T. Q. Collins, H. Patel and D. R. Berger, "Serine Oxidation via Glycine Cleavage (SOGC) Continues its Emergence as a Hallmark of Defective Mitochondria", *Cell Metabolism,* 23, pp. 635-648, 2016.
24. C. Feng-Song, Z. Ya-Ru, S. Hong-Cai, A. Zong-Hua, A. Su-Yi, Z. Su-Yi and W. Ju-Yong, "Deuterium-depleted water inhibits human lung carcinoma cell growth by apoptosis", *Experimental and Therapeutic Medicine,* 1, pp. 277-283, 2010.
25. W. Bild, I. Stefanescu, I. Haulica, C. Lupusoru, G. Titescu, R. Iliescu and V. Nastasa, "W. Bild, I. Stefanescu, I. Haulica, C. Lupusoru Research concerning the radioprotective and immunostimulating effects of deuterium-depleted water", *Rom. J. Physiol.,* 36, pp. 205-218, 1999.
26. I. Siniak, V. S. Turusov, A. I. Grigor'ev, D. G. Zaridze, V. B. Gaidadymov, E. I. Gus'kova, E. E. Antoshina and L. S. Trukhanova, "Consideration of the deuterium-free water supply to an expedition to Mars", *Aviakosm Ekolog Med.,* 37, pp. 60-63, 2003.
27. I. Paduraru, L. Jerca, A. Berbec, W. Wild, C. Lupusoru, I. Haulica, O. Paduraru and O. Jerca, "Deuterium depleted water effects over some oxidative stress parameters", *Roum. Biotech. Lett.,* 5, pp. 273-278, 2000.
28. V. S. Tyrysov, I. Siniak, E. E. Antoshina, L. S. Trykhanov and T. G. Gor'kova, "The effect of preliminary administration of water with reduced deuterium content on the growth of transplantable tumors in mice", *Vopr. Onkol.,* 52, pp. 59-62, 2006.
29. . H. Wang, C. Liu, W. Fang and H. Yang, "Research progress of the inhibitory effect of deuterium-depleted water on cancers", *J. South Med Univ,* 32, 2012.
30. Y. Kamal and K. Lida, "Deuterium Depleted Water Inhibits the Proliferation of Human MCF7 Breast Cancer Cell Lines by Inducing Cell Cycle Arrest", *Nutrition and Cancer,* 71, pp. 1019-1029, 2019.
31. C. J. Collins and N. S. Bowman, Isotope Effects in Chemical Reactions, Van Nostrand Reinhold: New York, 1971.
32. P. W. Rundel, J. R. Ehleringer and K. A. Nagy, Stable Isotopes in Ecological Research, New York: Springer, 1988.

33. G. Jancsó, Isotope Effects. In: Handbook of Nuclear Chemistry, Dordrecht, Netherlands: Kluwer Academic Publishers, 2003, pp. 85-116.
34. International Atomic Energy Agency, Statistical Treatment of Data on Environmental Isotopes in Precipitation: Technical Report Series, Vienna, 1992, pp. 784-.
35. A. T. Burcin Alev Tuzuner and Y. Aysen, "Is it Possible to Prepare Deuterium Depleted Water at Home?", *Clinical and Experimental Health Sciences,* 28, pp. 226–227, 2018.
36. J. J. Katz and H. L. Crespi, Isotope Effects in Biological Systems. In: Isotope Effects in Chemical Reactions, New York: Van Nostrand Reinhold, 1971, pp. 286-363.
37. D. M. Czajka, A. J. Finkel, C. S. Fischer and J. J. Katz, "Physiological effects of deuterium on dogs", *Am. J. Physiol.,* 201, p. 357, 1961.
38. H. Ziegler, C. B. Osmond, W. Stichler and P. Trimborn, "Hydrogen isotope discrimination in higher plants: Correlations with photosynthetic pathway and environment", *Planta,* 128, pp. 85–92, 1976.
39. O. L. Sternberg, J. M. Deniro and B. H. Johnson, "Isotope Ratios of Cellulose from Plants Having Different Photosynthetic Pathways", *Plant Physiol,* 74, pp. 557–561, 1984.
40. F. E. Marilyn, F. Estep and C. H. Thomas, "Stable Hydrogen Isotope Fractionations during Autotrophic and Mixotrophic Growth of Microalgae", *Plant Physiology,* 67, pp. 474-477, 1981.
41. J. R. Richard, J. Robins, B. Isabelle, D. Jia-Rong, G. S´ebastien Guiet, P. S´ebastien Pionnier and Z. Ben-Li, "Measurement of 2H distribution in natural products by quantitative 2H NMR: An approach to understanding metabolism and enzyme mechanism?", *Phytochemistry Reviews,* 2, pp. 87-102, 2003.
42. R. J. Robins, G. S. Remaud and I. Billault, "Robins, R. J., Remaud, G. S. & Billault, I. Natural mechanisms by which deuterium depletion occurs in specific positions in metabolites", *Eur. Chem. Bull. 1(1), 39–40 (2012).,* 1, pp. 39–40, 2012.

43. Z. Youping, Z. Benli, S. W. Hilary, G. Kliti, H. H. Charles, G. Arthur, E. K. Zachary and D. F. Graham, "On the contributions of photorespiration and compartmentation to the contrasting intramolecular 2H profiles of C3 and C4 plants sugar", *Phytochemistry,* 145, pp. 197-206, 2017.
44. G. Somlyai, I. Somlyai, I. Fórizs, G. Czuppon, A. Papp and M. Molnár, "Effect of systemic subnormal deuterium level on metabolic syndrome related and other blood parameters in humans: A preliminary study", *Molecules,* 25, p. 1376, 2020.
45. A. Kotyk, M. Dvorakova and J. Koryta, „Deuterons cannot replace protons in active transport processes in yeast", *FEBS Letters,* 264, pp. 203-205, 1990.
46. G. Török, M. Csík, A. Pintér and A. Surján, Török, G. "A táptalajok „normálistól" eltérő deutérium koncentrációjának hatása a baktériumok szaporodására and mutagenezisére", *Egészségtudomány,* 44, pp. 331-338, 2000.
47. J. Pouyssegur, J. C. Chambard, A. Franchi, S. Paris and E. Van Obberghen-Schilling, "Growth factor activation of an amiloride-sensitive Na⁺/H⁺ exchange system in quiescent fibroblasts: coupling to ribosomal protein S6 phosphorylation", *Proc. Natl. Acad. Sci.,* 79, pp. 3935-3939, 1982.
48. J. Pouyssegur, C. Sardet, A. Franchi, G. L'Allemain and S. A. Paris, "A specific mutation abolishing Na⁺/H⁺ antiport activity in hamster fibroblasts precludes growth at neutral and acidic pH", *Proc. Natl. Acad. Sci.,* kötet 81, pp. 4833–4837, 1984.
49. R. Perona and R. Serrano, "Increased pH and tumorigenicity of fibroblasts expressing a yeast proton pump", *Nature,* 334, pp. 438--440, 1988.
50. O. Warburg, "On the origin of cancer cells", *Science,* 123, pp. 309-314, 1956.
51. R. E. Harris, "Cyclooxygenase-2 (cox-2) and the inflammogenesis of cancer", *Subcell Biochem,* 42, pp. 93-126, 2007.
52. G. Somlyai, Győzzük le a rákot, Budapest: AKGA Junior Kiadó, 2000.

53. Z. Gyöngyi and G. Somlyai, "Deuterium Depletion can Decrease the Expression of c-myc, Ha-Ras and p53 Gene in Carcinogen-Treated Mice", *In vivo*, 14, pp. 437-440, 2000.
54. Z. Gyöngyi, F. Budán, I. Szabó, I. Ember, I. Kiss, I. Ember, I. Kiss, K. Krempels, I. Somlyai and G. Somlyai, "Deuterium Depleted Water Effects on Survival of Lung Cancer Patients and Expression of Kras, Bcl2 and Myc Genes in Mouse Lung", *Nutrition and Cancer*, 65, pp. 240-246, 2013.
55. K. G. Geiss, E. R. Bumgarner, B. Birditt, T. Dahl, N. Dowidar, L. D. Dunaway, L. ....Hood and K. Dimitrov, "Direct multiplexed measurement of gene expression with color-coded base pairs", *Nature Biotechnology*, 26, pp. 317–325, 2008.
56. G. Somlyai, I. Somlyai, Z. Gyöngyi and G. L. Boros, "Effects of deuterium on cell growth, gene expression, survival and relapse rates of cancer patients" in *4th International Congress on Deuterium Depletion*, Budapest, 2019.
57. T. Strekalova, M. Evans, A. Chernopiatko, Y. Couch, J. Costa-Nunes, R. Cespuglio, L. Chesson, J. Vignisse, H. W. Steinbusch, D. C. Anthony, I. Pomytkin and K. P. Lesch, "Deuterium content of water increases depression susceptibility: The potential role of a serotonin-related mechanism", *Behavioural Brain Research*, 277, pp. 237-244, 2015.
58. C. Mladin, A. Ciobica, R. Lefter, A. Popescu and W. Bild, "Deuterium-depleted water has stimulating effects on long-term memory in rats", *Neuroscience Letters*, 583, pp. 154-158, 2014.
59. D. S. Ávila, G. Somlyai, I. Somlyai and M. Aschner, "Anti-aging effects of deuterium depletion on Mn-induced toxicity in a C. elegans model", *Toxicology Letters*, 211, pp. 319-324, 2012.
60. B. Aditya, M. T. Imad, R. J. Cerhan, K. A. Sood, J. P. Limburg, J. P. Erwin and M. V. Montori, "Efficacy of Antioxidant Supplementation in Reducing Primary Cancer Incidence and Mortality: Systematic Review and Meta-analysis", *Mayo Clin. Proc.*, 83, pp. 23-34, 2008.
61. Z. Xuepei Zhang, G. Massimiliano, A. Chernobrovkin and A. R. Zubarev, "Anticancer effect of deuterium depleted water - redox disbalance leads to oxidative stress", *Molecular & Cellular Proteomics*, 18, pp. 2373–2387, 2019.

62. W. Yongfu, Q. Dongyun, Y. Huiling, W. Wenya, X. Jifei, Z. Le and F. Hui, "Neuroprotective Effects of Deuterium-Depleted Water (DDW) Against H2O2-Induced Oxidative Stress in Differentiated PC12 Cells Through the PI3K/Akt Signaling Pathway", *Neurochemical Research*, 45, pp. 1034-1044, 2020.
63. G. Somlyai, M. Molnár, I. Somlyai, I. Fórizs, G. Czuppon, K. Balogh, O. Abonyi and K. Krempels, "A szervezet szubnormális deutériumszintjének kedvező élettani hatása a glükózintoleranciára, valamint a szérum HDL- and és Na$^+$- koncentrációra", *Egészségtudomány*, LVIII, 2014.
64. G. Somlyai, I. L. Nagy, L. Puskás, G. Fábián, Z. Gyöngyi, K. Krempels, I. Somlyai, G. Laskay, G. Jancsó, G. Jákli, A. Kovács, I. Guller, D. Avila and M. Ashner, "Hydrogen membrane transport activity coupled with changing deuterium/hydrogen ratio may be a key proliferation signal for the cells" in *4th Annual Meeting of the International Society of Proton Dynamics in Cancer, 10 Oct - 12 Oct*, Garching, Germany, 2013.
65. K. Krempels, I. Somlyai, Z. Gyöngyi, I. Ember, K. Balog, O. Abonyi and G. Somlyai, "A retrospective study of survival in breast cancer patients undergoing deuterium depletion in addition to conventional therapies", *Journal of Cancer Research & Therapy*, 1, pp. 194–200, 2013.
66. A. Kovács, I. Guller, K. Krempels, I. Somlyai, I. Jánosi, Z. Gyöngyi, I. Szabó, I. Ember and G. Somlyai, "Deuterium Depletion May Delay the Progression of Prostate Cancer", *Journal of Cancer Therapy*, 2, pp. 548-556, 2011.
67. L. G. Boros, E. J. Meuillet, I. Somlyai, G. Jancsó, G. Jákli, K. Krempels, L. G. Puskás, L. Nagy, M. Molnár, K. R. Laderoute, P. A. Thompson and G. Somlyai, "Fumarate hydratase and deuterium depletion control oncogenesis via NADPH-dependent reductive synthesis" in *AACR 2014 - Annual Meeting, April 5–9, San Diego, CA, USA*, 2014.
68. K. Krempels, I. Somlyai and G. Somlyai, "A retrospective evaluation of the effects of deuterium depleted water consumption on four patients with brain metastases from lung cancer", *Integrative Cancer Therapies*, 7, pp. 172–181, 2008.

69. R. B. Corcoran and B. A. Chabner, "Application of cell-free DNA analysis to cancer treatment", *N Engl J. Med.*, 379, pp. 1754-1765, 2018.
70. C. Abbosh, N. J. Birkbak, G. A. Wilson and C. Swanton, "Phylogenetic ctDNA analysis depicts early-stage lung cancer evolution", *Nature,* 545, pp. 446-451, 2017.
71. T. Minamoto, E. Wada and I. Shimizu, "A new method for random mutagenesis by error-prone polymerase chain reaction using heavy water", *Journal of Biotechnology 157:71-74.*, 157, pp. 71-74, 2011.
72. J. Boren, M. Cascante, S. Marin, B. Comín-Anduix, J. J. Centelles, S. Lim, S. Bassilian, S. Ahmed, W. N. Lee and L. G. Boros, "Gleevec (STI571) influences metabolic enzyme activities and glucose carbon flow toward nucleic acid and fatty acid synthesis in myeloid tumor cells", *J Biol Chem.*, pp. 37747-37753, 2001.
73. N. Mut-Salud, P. Juan Álvarez, J. Manuel Garrido, E. Carrasco, A. Aránega and F. Rodrígez-Serrano, "Antioxidant Intake and Antitumor Therapy: Toward Nutritional Recommendations for Optimal Results", *Oxid Med Cell Longev.*, p. 6719534, 2016.
74. M. Molnár, K. Horváth, T. Dankó and G. Somlyai, "Effect of deuterium oxide (D2O) content of drinking water on glucose metabolism on STZ-induced diabetic rats" in *Proceedings of the 7th International Conference Functional Foods in the Prevention and Management of Metabolic Syndrome*, 2010.
75. I. Györe and G. Somlyai, "Csökkentett deutérium tartalmú ivóvíz hatása a teljesítőképességre sportolóknál", *Sportorvosi Szemle,* 46, pp. 27-38, 2005.

Acknowledgments

Over the past decades, I have had a lot of help, support, and encouragement from many people. If I wanted to thank them, I would need to dedicate an entire chapter in my book for the acknowledgments. Looking back on the early days, in retrospect, we can say that we were undertaking a "mission impossible." In the early 1990s, all we had was the results of merely half a year's research and the ambitious goal to develop a more effective anti-cancer drug with the fewest side effects. We founded HYD LLC for Cancer Research and Drug Development with the first scientific results to achieve this goal, knowing that the cost of having a drug registered costs approximately $920,000,000–$1,836,000,000 (HUF 300–600 billion in Hungary). Our company had an available capital of only $1,500 (HUF 500,000) at the beginning. In the decades since then, the whole world has taken notice of our research results and related drug development. Today, many leading universities and research institutes are investigating the biological effects of deuterium, its physiological role, and the therapeutic effects of deuterium depletion for several indications. We registered the world's first reduced deuterium-depleted veterinary drug (Vetera-DDW-25®) and more than two decades ago, we launched the world leader Preventa product range of deuterium-depleted drinking water, protected by international trademarks, and sold in more than fifty countries around the world. We have successfully implemented a substantial pharmaceutical investment, the first production line that works in compliance with international standards to produce deuterium-depleted water. Today, there are thousands of people alive or have lived considerably longer thanks to the consumption of deuterium-depleted water.

Support and help from hundreds of people were needed for our company, HYD LLC for Cancer Research and Drug Development, to come this far. First and foremost, I would like to thank my wife, the biologist Ildikó Somlyai, whose faith, perseverance, hard work, advice, and professionalism contributed a great deal to the success of our company and helped solve our challenges and tasks. She also played a key role in shaping the contents of my book. The ideas that I have written down retained their final form through her editing and proofreading work. I also owe thanks to the members of my family, my daughters Dóra and Szilvia, who shared with us the burden of our work and were happy for our successes. My daughter Dóra, with her painstaking and accurate work, has been an integral part of our team for years. Szilvi, my other daughter, a communication-professional-turned yoga instructor, and nutrition coach, has been a supportive and helpful advisor. I am also indebted to my parents and teachers who started my career and raised my interest in the secrets of nature and science.

I would like to thank all the previous and current employees, owners, and investors of HYD LLC for Cancer Research and Drug Development. Our company has faced and coped with several challenges and difficulties, but our colleagues have acted as an excellent team in every situation.

I also owe special thanks to all the partners of HYD LLC. We shared success and difficulties in the past decades, and our partners shared our joys in the successes and sympathized during the hard times.

We have received considerable support from fellow researchers, home and abroad. They acknowledged the importance of our first publication back in 1993 and the ones that followed it, joining our research of deuterium depletion. I especially want to thank Gábor Jancsó and György Jákli of the Nuclear Energy Research Institute who were the first to support me. Without their professional knowledge and help, I couldn't have started my research on deuterium depletion. I would also like to thank the work of professors László G. Boros (UCLA) and Roman Zubarev (Karolinska Institute, Stockholm). Both had a substantial contribution to exploring the mechanism of deuterium depletion. Valentin I. Lobyshev (Moscow State University) and Stephan S. Dzhimak (Kuban State University, Krasnodar) have contributed to the research with many results. Scientific cooperation with the following researchers in Hungary has also contributed greatly to the success of our research: Gábor Laskay (†) (University of Szeged), Miklós Molnár (Semmelweis University), László Puskás (Biological Research Center, Szeged) and Zoltán Gyöngyi (University of Pécs Medical School). Thanks to the open-mindedness and work ethic of researchers in the field, over a hundred international scientific papers have been published on this topic up until 2020 and still counting.

Looking back on the past thirty years, we can discover the same principles in the field of deuterium depletion research as in the rest of the world. One group of people, utilizing the best of their knowledge, are trying to look for solutions for the problems of our modern times. Others, sometimes abusing their power, question our efforts and results without any professionalism. This is why we appreciate so much the faith, trust, and work of those who do their jobs driven by scientific curiosity and are devoid of prejudice. Much appreciated is the helpful support of those without whom the research of deuterium depletion could not have taken its place in science and without whom the subject could not have gained international recognition. I am confident that people who are eager to change things will achieve their goals. I want our supporters and partners to share our success. Every life saved, every life extended, preserved or health restored is a testament of their work and contribution, worthy of the gratitude and appreciation of every one of us.

Gábor Somlyai PhD

Printed by publishdrive

	Lite water	Regular water
1st morning coffee	—	250 cc
2nd morning coffee	250	—
Coffee together	300 cc	—
Afternoon coffee at home	300 cc	100 cc
Water for pm pills	100 cc	60 cc

650
at home

300 at my place

650 cc
× 30
———
19,500 cc

39 bottles per month

$450/month

Me –
am coffee – 300 cc
coffee together – 500 cc